CLIMATE CHANGE AND ITS CAUSES, EFFECTS AND PREDICTION

POLICY OPTION ISSUES FOR CO_2 EMISSIONS

CLIMATE CHANGE AND ITS CAUSES, EFFECTS AND PREDICTION

Global Climate Change
Horace M. Karling
(Editor)
2001. ISBN: 1-56072-999-6

Global Climate Change Revisited
Harace B. Karling
(Editor)
2007. ISBN: 1-59454-039-X

Climate Change Research Progress
Lawrence N. Peretz
(Editor)
2008. ISBN: 1-60021-998-5

Climate Change: Financial Risks
United States Government Accountability Office
2008. ISBN: 978-1-60456-488-4

Post-Kyoto: Designing the Next International Climate Change Protocol
Matthew Clarke
2008. ISBN: 978-1-60456-840-0
2008. ISBN: 978-1-61668-101-2 (E-book)

The Effects of Climate Change on Agriculture, Land Resources, Biodiversity in the United States
Peter Backlund, Anthony Janeto and David Schimel
2009. ISBN: 978-1-60456-989-6

Economics of Policy Options to Address Climate Change
Gregory N. Bartos
2009. ISBN: 978-1-60692-116-6

Thresholds of Climate Change in Ecosystems
Vicente Orostegui
(Editor)
2009. ISBN: 978-1-60741-487-2

Climate Variability, Modeling Tools and Agricultural Decision-Making
Angel Utset
(Editor)
2009. ISBN: 978-1-60692-703-8
2009. ISBN: 978-1-60876-791-5 (E-book)

Emissions Trading: Lessons Learned from the European Union and Kyoto Protocol Climate Change Programs
Ervin Nagy and Gisella Varga
(Editors)
2009. ISBN: 978-1-60741-194-9

Constructing Climate Change Legislation: Background and Issues
Gerald P. Overhauser
(Editor)
2009. ISBN: 978-1-60692-986-5

Designing Greenhouse Gas Reduction and Regulatory Systems
Sonja Enden (Editor)
2009. ISBM: 978-1-60741-195-6

Global Climate Change: International Perspectives and Responses
Elias D'Angelo (Editor)
2009. ISBN: 978-1-60741-233-5

Disputing Global Warming
Anton Horvath and Boris Molnar (Editors)
2009. ISBN: 978-1-60741-235-9
2009. ISBN: 978-1-60876-503-4 (E-book)

Focus on Climate Change and Health
Viroj Wiwanitkit
2009. ISBN: 978-1-60741-247-2

Land Use and Climate Change
Suresh C. Rai
2009. ISBN: 978-1-60741-362-2

Impacts of Climate Change on Transportation and Infrastructure - A Gulf Coast Study
Iason Pavlopoulos (Editor)
2009. ISBN: 978-1-60741-424-7

Coastal Sensitivity to Sea Level Rise - Focusing on the Mid-Atlantic Region
Melvin C. Urajner (Editor)
2009. ISBN: 978-1-60741-440-7

Lightning in the Tropics: From a Source of Fire to a Monitoring System of Climatic Changes
Osmar Pinto, Jr.
2009. ISBN: 978-1-60741-764-4

Carbon Capture and Greenhouse Gases
Imrus Juhász and Gyorgy Halász (Editors)
2010. ISBN: 978-1-60692-089-3

The Science of Climate Change and Policy Implications
Kyle S. Hartzell (Editor)
2010. ISBN: 978-1-60741-448-3

Effects of Climate Change on Energy Production and Use in the U.S.
Alrik M. Solberg (Editor)
2010. ISBN: 978-1-60741-426-1

Policy Option Issues for CO2 Emissions
Nikolaus Vogler (Editor)
2010. ISBN: 978-1-60741-381-3

The Role of Agriculture in Carbon Capture and Climate Change
Arvid Bjurstrom (Editor)
2010. ISBN: 978-1-60741-445-2

Carbon Sequestration: Methods, Modeling and Impacts
Elke Hoch and Siegbert Grunwald (Editors)
2010. ISBN: 978-1-60741-498-8

Climate Change Litigation and Law
Jean-François Masson (Editor)
2010. ISBN: 978-1-60876-089-3

Carbon Tax and Cap-and-Trade Tools: Market-Based Approaches for Controlling Greenhouse Gases
Nelson E. Burney (Editor)
2010. ISBN: 978-1-60876-137-1

Global Change and Forestry: Economic and Policy Impacts and Responses
Jianbang Gan, Stephen Grado and Ian A. Munn (Editors)
2010. ISBN: 978-1-60876-262-0

Carbon Offsets: Examining their Role in Greenhouse Gas Reduction
Karen T. Morningstar (Editor)
2010. ISBN: 978-1-60741-444-5

Climate Change Measures and Trade Considerations
George R. Fried (Editor)
2010. ISBN: 978-1-60876-756-4

Assessing Climate Change Impacts on the United States
Alex B. McNeill (Editor)
2010. ISBN: 978-1-60876-160-9
2010. ISBN: 978-1-61668-537-9 (E-book)

Responding to Impacts of Climate Change on Water Resources
Zachary E. Quinn (Editor)
2010. ISBN: 978-1-60741-992-1
2010. ISBN: 978-1-61728-095-5 (E-book)

Climate Change: Fundamental Issues and Policy Tools
Elise M. Farrugia (Editor)
2010. ISBN: 978-1-60741-997-6
2010. ISBN: 978-1-61728-052-8 (E-book)

Effects of Climate Change on Aquatic Invasive Species
Sofia A. Contreras (Editor)
2010. ISBN: 978-1-61728-005-4
2010. ISBN: 978-1-61728-251-5 (E-book)

Paleoecology of Peatlands: Quaternary Climate Reconstruction from Hungary*
Gusztáv Jakab, Pál Sümegi and Erzsébet Szurdoki
2010. ISBN: 978-1-61728-220-1
2010. ISBN: 978-1-61728-792-3 (E-book)

CLIMATE CHANGE AND ITS CAUSES, EFFECTS AND PREDICTION

POLICY OPTION ISSUES FOR CO_2 EMISSIONS

NIKOLAUS VOGLER
EDITOR

Nova Science Publishers, Inc.
New York

For permission to use material from this book please contact us:
Telephone 631-231-7269; Fax 631-231-8175
Web Site: http://www.novapublishers.com

LIBRARY OF CONGRESS CATALOGING-IN-PUBLICATION DATA

Policy option issues for CO2 emissions / editor, Nikolaus Vogler.
p. cm.
Includes index.
ISBN 978-1-60741-381-3
1. Carbon dioxide mitigation--Government policy--United States. 2. Environmental policy--Economic aspects--United States. 3. Greenhouse gas mitigation--Government policy--United States. 4. Carbon taxes--United States. I. Vogler, Nikolaus.
HC110.A4P65 2009
363.738'745610973--dc22

2009041009

Published by Nova Science Publishers, Inc. ✢ New York

CONTENTS

PREFACE

Global climate change is one of the nation's most significant long-term policy challenges. Human activities are producing increasingly large quantities of greenhouse gases, particularly carbon dioxide (CO2), which accumulate in the atmosphere and create costly changes in regional climates throughout the world. This book emphasizes how the most efficient approaches to reducing emissions involve giving businesses and individuals an incentive to curb activities that produce CO2 emissions. This book consists of public documents which have been located, gathered, combined, reformatted, and enhanced with a subject index, selectively edited and bound to provide easy access.

Chapter 1 - Human activities are producing increasingly large quantities of greenhouse gases, particularly carbon dioxide (CO_2), and their accumulation in the atmosphere is expected to affect the climate throughout the world. This Congressional Budget Office issue brief examines the role of passenger vehicles (cars and light trucks) in the U.S. effort to curb those emissions. In particular, the brief looks at how putting a price on CO_2 emissions—for example, through a cap-and-trade system—would affect gasoline prices and, as a conse-quence, vehicle emissions.

Charging a price for CO_2 emissions would raise the price of gasoline, but that increase—and the resulting decrease in vehicle emissions—would be relatively small. Most of the reduction in CO_2 emissions would occur in other sectors.

The initial impact on vehicle emissions would be particularly small: People could drive less and at slower speeds, and some could switch to public transit, but in the short run they would have few other alternatives. Over time, consumers could respond to higher gasoline prices by buying more fuel-efficient vehicles and reducing their commuting distance when an opportunity arises. Substantial increases in gasoline prices in recent years have triggered measurable responses of both types. But a CO_2 price high enough to induce sizable reductions from other sources of emissions would have only a small effect on vehicle emissions of CO_2. Recent changes to the automobile fuel economy standards—greatly increasing their stringency—will result in a substantial decline in vehicle emissions whether gasoline prices increase or not.

Global climate change is among the most serious long-term challenges facing the nation. The accumulation of greenhouse gases in the atmosphere could have serious and costly effects throughout the world. Although the magnitude of those effects remains highly uncertain, there is growing recognition of the risk that it may be extensive and perhaps catastrophic.

Reducing greenhouse-gas emissions would lower the economic and human health risks associated with a changing climate. The primary greenhouse gas is carbon dioxide (CO_2), and according to the Environmental Protection Agency (EPA), about 20 percent of total U.S. emissions of CO_2 are from passenger vehicles (cars and light trucks). Those emissions are directly related to the amount of gasoline a vehicle uses, which in turn depends on the number of miles the vehicle is driven and on its fuel economy.

For many households, the choices of which car to drive and how much to drive it are among the most visible ways in which individuals contribute to climate change. Yet research suggests that policies to reduce greenhouse-gas emissions by setting a price on them (through a cap-and-trade system or a carbon tax, for example) would have relatively little effect on vehicle emissions. Instead, most of the reductions would come from other sources— particularly electric power generators—from which emissions might be reduced at lower cost.

A cap-and-trade system or a carbon tax would raise the price of gasoline, encouraging consumers to drive less and to buy vehicles that are more fuel efficient, but the effects on the price of gasoline and on consumers' choice of vehicles and driving behavior would be modest under most policy proposals. For example, despite the recent dramatic rise in gasoline prices—substantially more than would occur under the types of climate policy being discussed—the decline in gasoline consumption and, correspondingly, in vehicle emissions has been relatively small.

A study by the Congressional Budget Office (CBO) found that the rise in gasoline prices between 2003 and 2007 (from \$1.50 to more than \$3.00 per gallon) caused only a small decline in the amount of driving; a slight reduction in vehicle speeds on uncongested freeways; a moderate increase in the purchase of cars relative to light trucks, such as sport–utility vehicles (SUVs) and minivans; and somewhat better average fuel economy for new cars and light trucks.

Furthermore, imposing a price on CO_2 emissions is unlikely to cause the passenger-vehicle fleet to become more fuel efficient because the recently revised corporate average fuel economy (CAFE) standards already require greater improvements in fuel economy than a CO_2 price would achieve. Those standards will result in a decline in CO_2 emissions irrespective of whether those emissions are priced. Thus, reductions in vehicle emissions would constitute only a small fraction of the total reduction in CO_2—probably less than 5 percent—that would occur under a policy of pricing CO_2 emissions.

This issue brief updates the findings from CBO's earlier study, describing how consumers responded as gasoline prices continued to climb, to more than $4 per gallon by May 2008, where they remained for much of the summer. The brief then looks at how pricing CO_2 emissions would affect passenger vehicles and driving behavior, and it assesses the potential reductions in vehicles' CO_2 emissions that would result from such a pricing policy.

Chapter 2 - This chapter is edited and excerpted testimony by Peter R. Orszag before the Committee on Ways and Means on September 18, 2008.

Chapter 3 - Global climate change is one of the nation's most significant long-term policy challenges. Human activities are producing increasingly large quantities of greenhouse gases, particularly carbon dioxide (CO_2), which accumulate in the atmosphere and create costly changes in regional climates throughout the world. The magnitude of such damage remains highly uncertain, but there is growing recognition that some degree of risk exists for the damage to be large and perhaps even catastrophic. Reducing greenhouse-gas emissions would be beneficial in limiting the degree of damage associated with climate change. However, decreasing those emissions would also impose costs on the economy—in the case of CO_2, because much economic activity is based on fossil fuels, which release carbon in the form of carbon dioxide when they are burned. Most analyses suggest that a carefully designed program to begin lowering CO_2 emissions would produce greater benefits than costs.

The most efficient approaches to reducing emissions involve giving businesses and individuals an incentive to curb activities that produce CO_2 emissions, rather than adopting a "command and control" approach in which the government would mandate how much individual entities could emit or what technologies they should use. Incentive-based policies include a tax on emissions, a cap on the total annual level of emissions combined with a system of tradable emission allowances, and a modified cap-and-trade program that includes features to constrain the cost of emission reductions that would be undertaken in an effort to meet the cap. In this study, the Congressional Budget Office (CBO) compares these incentive-based approaches, focusing on three key criteria:

- Efficiency in maintaining a balance between the uncertain benefits and costs of reducing CO_2 emissions,
- Ease or difficulty of implementation, and
- Possible interactions with other countries' policies for curbing CO_2—that is, the potential to ensure that U.S. and foreign policies produce similar incentives to cut emissions inside and outside the United States.

Other criteria could be of interest to policymakers in determining how best to address concerns about climate change. For example, the efficiency criterion addresses how well policies might function to minimize the cost of reducing emissions over a period of several decades; however, policymakers may choose to place more emphasis on providing certainty about the amount of emissions at specific points in time. Similarly, policymakers may also wish to focus on how different policy designs affect different segments of society.

Chapter 4 - Several important human activities—most notably the worldwide burning of coal, oil, and natural gas—are gradually increasing the concentrations of carbon dioxide and other greenhouse gases in the atmosphere and, in the view of many climate scientists, are gradually warming the global climate. That warming, and any long-term damage that might result from it, could be reduced by restraining the growth of greenhouse gas emissions and ultimately limiting them to a level that stabilized atmospheric concentrations.

The magnitude of warming and the damages that might result are highly uncertain, in part because they depend on the amount of emissions that will occur both now and in the future, how the global climate system will respond to rising concentrations of greenhouse gases in the atmosphere, and how changes in climate will affect the health of human and natural systems. The costs of restraining emissions are also highly uncertain, in part because they will depend on the development of new technologies.[1] From an economic point of view, the challenge to policy-makers is to implement policies that balance the uncertain costs of restraining emissions against the benefits of avoiding uncertain damages from global warming or that minimize the cost of achieving a target level of concentrations or level of annual emissions.

Researchers have studied the relative efficacy—as well as the appropriate timing—of various policies that might discourage emissions of carbon dioxide (referred to as carbon emissions in the rest of this paper), which makes up the vast majority of greenhouse gases, and restrain the growth of its atmospheric concentration. This paper presents qualitative findings from that research, which are largely independent of any particular estimate of the costs or benefits of reducing emissions. The paper's conclusions are summarized below.

In: Policy Option Issues for CO_2 Emissions ISBN: 978-1-60741-381-3
Editors: Nikolaus Vogler

Chapter 1

CLIMATE CHANGE, CO_2 EMISSIONS & PASSENGER VEHICLES

Congressional Budget Office

SUMMARY

Human activities are producing increasingly large quantities of greenhouse gases, particularly carbon dioxide (CO_2), and their accumulation in the atmosphere is expected to affect the climate throughout the world. This Congressional Budget Office issue brief examines the role of passenger vehicles (cars and light trucks) in the U.S. effort to curb those emissions. In particular, the brief looks at how putting a price on CO_2 emissions—for example, through a cap-and-trade system—would affect gasoline prices and, as a conse-quence, vehicle emissions.

Charging a price for CO_2 emissions would raise the price of gasoline, but that increase—and the resulting decrease in vehicle emissions—would be relatively small. Most of the reduction in CO_2 emissions would occur in other sectors.

The initial impact on vehicle emissions would be particularly small: People could drive less and at slower speeds, and some could switch to public transit, but in the short run they would have few other alternatives. Over time, consumers could respond to higher gasoline prices by buying more fuel-efficient vehicles and reducing their commuting distance when an opportunity arises. Substantial increases in gasoline prices in recent years have triggered measurable responses of both types. But a CO_2 price high enough to induce sizable reductions from other sources of emissions would have only a small effect on vehicle emissions of CO_2. Recent changes to the automobile fuel economy standards—greatly increasing their stringency—will result in a substantial decline in vehicle emissions whether gasoline prices increase or not.

Global climate change is among the most serious long-term challenges facing the nation. The accumulation of greenhouse gases in the atmosphere could have serious and costly effects throughout the world. Although the magnitude of those effects remains highly uncertain, there is growing recognition of the risk that it may be extensive and perhaps catastrophic.

Reducing greenhouse-gas emissions would lower the economic and human health risks associated with a changing climate. The primary greenhouse gas is carbon dioxide (CO_2), and according to the Environmental Protection Agency (EPA), about 20 percent of total U.S. emissions of CO_2 are from passenger vehicles (cars and light trucks). Those emissions are directly related to the amount of gasoline a vehicle uses, which in turn depends on the number of miles the vehicle is driven and on its fuel economy.

For many households, the choices of which car to drive and how much to drive it are among the most visible ways in which individuals contribute to climate change. Yet research suggests that policies to reduce greenhouse-gas emissions by setting a price on them (through a cap-and-trade system or a carbon tax, for example) would have relatively little effect on vehicle emissions. Instead, most of the reductions would come from other sources—particularly electric power generators—from which emissions might be reduced at lower cost.

A cap-and-trade system or a carbon tax would raise the price of gasoline, encouraging consumers to drive less and to buy vehicles that are more fuel efficient, but the effects on the price of gasoline and on consumers' choice of vehicles and driving behavior would be modest under most policy proposals. For example, despite the recent dramatic rise in gasoline prices—substantially more than would occur under the types of climate policy being discussed—the decline in gasoline consumption and, correspondingly, in vehicle emissions has been relatively small.

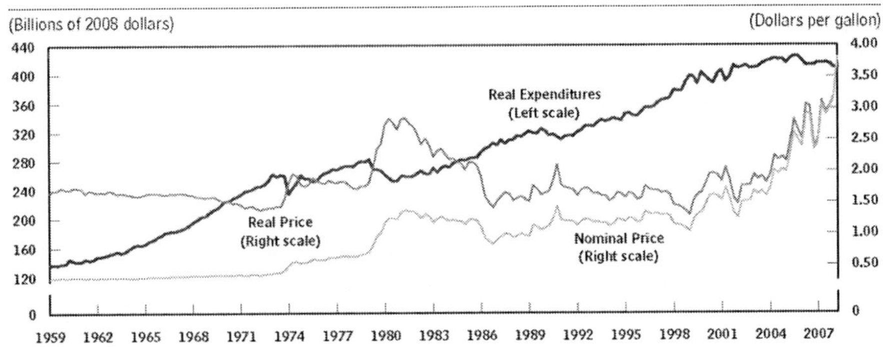

Source: Congressional Budget Office based on data from the Bureau of Economic Analysis and the Energy Information Administration.

Note: Consumer expenditures are for gasoline and motor oil through June 2008 (motor oil is about 1.5 percent of the total).

Consumer expenditures were adjusted for inflation by CBO using the Bureau of Economic Analysis's (BEA's) chained price index for gasoline and other motor fuel. Changes in expenditures therefore reflect changes in gallons consumed. Real (inflation-adjusted) gasoline prices were calculated by CBO using BEA's consumer price index for all urban consumers.

Figure 1. Personal Expenditures for Gasoline and the Average Price of Gasoline in the United States

A study by the Congressional Budget Office (CBO) found that the rise in gasoline prices between 2003 and 2007 (from \$1.50 to more than \$3.00 per gallon) caused only a small decline in the amount of driving; a slight reduction in vehicle speeds on uncongested freeways; a moderate increase in the purchase of cars relative to light trucks, such as sport–utility vehicles (SUVs) and minivans; and somewhat better average fuel economy for new cars and light trucks.[1]

Furthermore, imposing a price on CO_2 emissions is unlikely to cause the passenger-vehicle fleet to become more fuel efficient because the recently revised corporate average fuel economy (CAFE) standards already require greater improvements in fuel economy than a CO_2 price would achieve.[2] Those standards will result in a decline in CO_2 emissions irrespective of whether those emissions are priced. Thus, reductions in vehicle emissions would constitute only a small fraction of the total reduction in CO_2—probably less than 5 percent—that would occur under a policy of pricing CO_2 emissions.

This issue brief updates the findings from CBO's earlier study, describing how consumers responded as gasoline prices continued to climb, to more than

$4 per gallon by May 2008, where they remained for much of the summer.[3] The brief then looks at how pricing CO_2 emissions would affect passenger vehicles and driving behavior, and it assesses the potential reductions in vehicles' CO_2 emissions that would result from such a pricing policy.

GASOLINE PRICES, DRIVING BEHAVIOR, AND CHOICE OF VEHICLE

In response to increases in the price of gasoline, individuals can reduce their gasoline consumption by changing their driving behavior, the type of vehicle they drive, and eventually where they choose to live and work. In the short run, most of their adjustment is in the form of changed behavior, with only a modest effect on gasoline consumption. In the longer run, consumption becomes more sensitive to higher prices because motorists are able to respond in ways they cannot in the short run—particularly by choosing vehicles that get better gasoline mileage. Sustained high prices would eventually alter land-use patterns, as people began to seek homes and job locations that would reduce their commuting distance.

Table 1. Estimated Effects of a $2 Increase in the Price of Gasoline on Speeds on Uncongested Highways

(Miles per hour) Speed Percentile	Speed in 2004	Reduction in Speed
15th	65.4	1.4 to 2.4
Median	68.4	1.0 to 1.5
85th	70.8	0.6 to 0.75

Source: Congressional Budget Office based on data from the California Department of Transportation for January 2004 through April 2008.

Note: Sample includes six freeway locations in California: North-bound I-680 in San Ramon, east of Oakland; Northbound I-880 in San Jose; Westbound SR 60 in City of Industry (eastern Los Angeles County); Southbound I-405 in Westminster (Orange County); Eastbound SR 78 in San Marcos (northern San Diego County); Westbound I-8 in San Diego.

Driving Behavior

Although some important aspects of driving behavior— particularly the length of the commute to work—cannot be adjusted quickly in response to changes in the price of gasoline, motorists can quickly make other changes to save gasoline (and they can revert just as quickly if prices go down). They can drive more slowly, accelerate more gradually, take fewer discretionary trips, use shopping or recreation sites that are closer to home, or switch to other modes of transportation where possible.

In response to higher gasoline prices beginning in 2003, gasoline consumption began to grow more slowly in the United States and, eventually, to decline (see Figure 1).[4] With sharply higher prices in 2008, total miles of vehicle travel have been lower each month than in the corresponding month in 2007—a phenomenon not seen in the United States since 1979. Through June, motorists drove 2.8 percent fewer miles than in the first six months of 2007.[5] Moreover, the higher prices in 2008 have reinforced the decline in driving speeds on uncongested free-ways that was identified in CBO's January 2008 study, as well as the increase in ridership on public transit.

For this issue brief, CBO updated and expanded its analysis of driving speeds on uncongested freeways, using data from 2004 through April 2008 for six California freeway locations. Over that time, as the price of gasoline increased by \$2—to nearly \$4 per gallon—the median speed of freeway travel in uncongested conditions declined between 1.0 mile per hour (mph) and 1.5 mph (see Table 1). The resulting fuel savings are consistent with estimates of how prices affect gasoline consumption in the short run. According to those estimates, consumption tends to decline by about 0.6 percent for every 10 percent increase in the price of gasoline.[6] The decline in speed was somewhat greater for vehicles traveling at slower speeds, and it was smaller for vehicles moving at faster speeds.[7] That result is consistent with the notion that motorists' responses depend on how they value their time: If motorists who place a higher value on their time tend to drive faster than the median driver, they will also be less likely to slow down in response to higher gasoline prices, because they value saving time more than saving on fuel costs.[8] Similar logic could explain why the response is greater for vehicles traveling at below-median speeds.

Choices of New and Used Vehicles

In the long run, the response to higher gasoline prices is estimated to be about seven times greater than in the short run: A sustained increase in the price of gasoline would ultimately reduce consumption by about 4 percent for every 10 percent increase in price.[9] (Such an estimate may be less applicable to larger price increases, however, because there are practical limits to how much people can reduce their use of gasoline.) The larger response over time reflects consumers' greater ability to use less fuel by eventually making more dramatic changes than simply driving at slower speeds—in particular, by replacing their vehicles with ones that have greater fuel efficiency. With higher gasoline prices over the past few years, demand has shifted toward more fuel-efficient vehicles. If gasoline prices remain high, that shift is likely to continue.

Between 2003 and 2006, the average rated fuel economy of new cars and light trucks sold in the United States increased by about 1 mile per gallon (mpg), from 24.3 mpg to 25.2 mpg, according to EPA's calculations.

The average for 2007 was 25.7 mpg.[10] Based on monthly data for the first half of 2008, the average for the year will probably be higher. Those increases follow several decades in which average fuel economy remained steady or gradually declined (see Figure 2).

Factors other than gasoline prices have contributed to the increase in average fuel economy. The CAFE standard for light trucks was raised by 1.5 mpg between 2004 and 2007.[11] Yet the influence of higher gasoline prices is clear because average fuel economy for new cars also increased over that time, even though the CAFE standard for those vehicles did not change.

The most important factor in the overall increase in average fuel economy has been the substantial recent growth in the share of cars among all new passenger vehicles. In 1981, more than 80 percent of all new passenger vehicles were cars. Since then, as light trucks—first minivans, then SUVs—became more popular, the share of cars fell every year, bottoming out at less than 45 percent in 2004 (see Figure 3). In response to increases in the price of gasoline over the past several years, the market share for cars has rebounded, reaching a seasonally adjusted annual rate of 56 percent in May, June, and July 2008, while the average price of gasoline was above $4 per gallon.

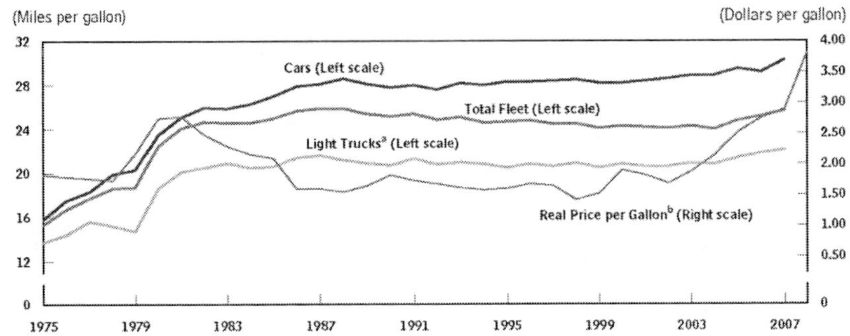

Source: Congressional Budget Office based on data from the Environmental Protection Agency.

Note: Data are for sales-weighted fuel economy. The Environmental Protection Agency determines a vehicle's fuel economy performance, either through its laboratory results or in test data submitted by the manufacturer, and the National Highway Traffic Safety Administration (NHTSA) determines compliance with corporate average fuel economy (CAFE) standards. NHTSA considers CAFE credits the automaker has earned, including those for hybrid and dual-fuel vehicles. NHTSA's CAFE data are similar to those illustrated here, although the averages are slightly higher because of the credits.

a. Includes sport–utility vehicles and minivans.

b. The real (inflation-adjusted) gasoline price for 2008 is the Energy Information Administration's estimated annual average price as of July 2008. Real (2008) prices were calculated by CBO using the Bureau of Economic Analysis's implicit price deflator.

Figure 2. Average Rated Fuel Economy for New U.S. Passenger Vehicles and the Real Price of Gasoline

That turnaround is all the more noteworthy because automakers have been raising prices more quickly for cars than for light trucks, according to CBO's analysis of two years of manufacturer's suggested retail prices (MSRPs).[12] For the nearly 200 vehicle models in that analysis, the average increase in MSRP for the identical model of car was 1.2 percent between the 2005 and 2006 model years, compared with only 0.3 percent for SUVs and minivans. That pricing pattern reflects the shift in consumer preferences toward smaller, more fuel-efficient vehicles as the price of gasoline first exceeded \$3 and then remained there for an extended time. Furthermore, within each category of vehicle, MSRPs have been rising more quickly for models with better fuel economy ratings.[13]

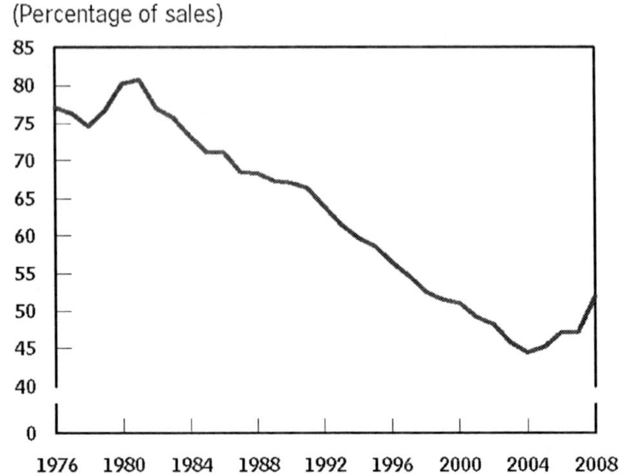

Source: Congressional Budget Office based on data from the Bureau of Economic Analysis.

Note: Market share for 2008 reflects the seasonally adjusted rate through July.

Figure 3. Share of Cars Among Sales of New Passenger Vehicles

IMPLICATIONS OF HIGHER GASOLINE PRICES FOR VEHICLE EMISSIONS

The adjustments that people have made in how (and how much) they drive and in the types of vehicles they are buying have been in response to larger increases in gasoline prices than would be likely to occur under any of the current proposals for pricing CO_2 emissions. The findings of this analysis, along with CBO's previous work and that of others to estimate the costs of proposed policies for dealing with climate change, provide a basis for estimating how gasoline prices and vehicles' CO_2 emissions would be affected by such policies.

CBO has estimated that a price in 2012 of $28 per metric ton of CO_2 (and other equivalent greenhouse gases) would lead to a reduction of about 10 percent in total U.S. emissions for that year compared with what would be expected if emissions were not priced.[14] That price per ton of CO2 emitted would add about 25 cents to the price of a gallon of gasoline—about a 6 percent increase if gasoline cost $4 per gallon.[15] In the short run, total gasoline consumption (and thus CO_2 emissions from vehicles) would remain essentially the same in response to such a small increase in price.[16] Over time, CO_2 emissions from vehicles would decline by around 2.5 percent, all else being equal—much less than from other sources given that the average reduction in emissions would be 10 percent—as consumers took the increase in the price of gasoline into account when replacing existing vehicles and revisiting decisions about where to live or work.

Several factors account for the relatively small influence that a price on CO_2 emissions would have on passenger vehicles and driving behavior. They include:

- The much smaller effect an emissions price would have on gasoline prices relative to the recent increase in those prices;[17] and
- The extent to which Americans have become dependent on automobile travel.

Furthermore, the volume of emissions in coming years will be heavily influenced by new, more stringent CAFE standards that will result in substantial gains in fuel economy over the next dozen years (see Box 1).[18] Correspondingly, pricing CO_2 emissions would not have any additional effect on fuel economy beyond what the CAFE standards already require, unless gasoline prices were much higher than they currently are.

A comparison of average vehicle fuel economy and gasoline prices in the United States with those in the European Union supports the conclusion that a very high CO_2 price would be necessary to significantly reduce vehicle emissions. In 2006, the average fuel economy for new passenger vehicles in the European Union was about 38 mpg.[19] Europe's higher fuel economy is due primarily to its much higher fuel taxes; the European Union has no mandatory standards for fuel economy.[20] Taxes on gasoline in Europe, levied by each country individually, vary between €0.51 and €0.57 per liter, or about $2.40 to $3.10 per gallon, depending on the exchange rate. Those taxes are about five to six times higher than the U.S. average of $0.47 per gallon.[21]

Tax differences are not the entire story, however. Cultural, historical, geographic, and infrastructural differences between the United States and Europe have also contributed to Europe's greater average fuel economy. The wide adoption of fuel-efficient diesel-powered cars in Europe may also have played a small role.[22] Thus, if gasoline prices in the United States equaled those in Europe, average fuel economy in the United States would probably approach the average for the European Union but would remain somewhat lower.

Box 1. CAFE Standards and Vehicles' CO_2 Emissions

Beginning in 2011, corporate average fuel economy (CAFE) standards will become more stringent and will vary for vehicles of different sizes. By law, 2020 model-year passenger vehicles (cars and light trucks) must average at least 35 miles per gallon (mpg) of gasoline, an increase of almost 10 mpg above the average for 2007 model-year vehicles.[1] The National Highway Transportation Safety Administration has the discretion to raise the standards above 35 mpg by 2020. Even at the statutory standard, new passenger vehicles will emit about 28 percent less carbon dioxide (CO_2) per mile. By 2035, when most pre-2020 passenger vehicles will have been retired, the new CAFE standards could be reducing total U.S. emissions of CO_2 by about 5 percent or more, depending on the number of registered vehicles, the price of gasoline, the rate of growth in CO_2 emissions elsewhere in the economy, and whether the CAFE standards are further tightened after 2020.

A number of factors could help minimize the cost of meeting tighter standards. Technological advances could expand the opportunities for saving fuel. Also, higher gasoline prices could encourage increased demand for fuel-efficient vehicles, which would make it easier for automakers to sell enough of those vehicles to comply with the standards.

The new fuel-economy standards could result in lower relative prices for vehicles that are more fuel efficient than other vehicles of similar size. Automakers have previously used such pricing practices as part of an overall strategy for complying with CAFE standards.[2] In influencing automakers' strategies for pricing vehicles, the standards may not only affect how vehicles are designed but may also provide consumers with financial incentives to buy vehicles that are more fuel efficient and disincentives to buy vehicles that have more power and better performance—attributes that are necessarily traded off against improved fuel economy.

Although a price on CO_2 emissions would increase demand for fuel economy, in the presence of a stringent CAFE standard that price would probably have little or no effect on average fuel economy. The CO_2 price, together with existing gasoline taxes, would strengthen the incentive for consumers to buy vehicles that are more fuel efficient, up to the point at which their additional cost per gallon of fuel saved—their expenditure on fuel-saving technologies (reflected in the vehicle's price) and the value of their forgone gains in performance—would equal the CO_2 price. (The location of that point depends on the amount of driving each consumer expects to do.) But on the margin, only one of the policies would actually boost fuel economy. Either the price of the CO_2 permit would be high enough to stimulate demand for fuel economy in excess of what the CAFE standard would require, or the standard would require fuel economy in excess of that demand.[3]

Although a cap-and-trade system would probably have little effect on fuel economy with stringent CAFE standards in place, as long as vehicles continued to run on gasoline (or on coal-generated electric power) a cap-and-trade system would further reduce vehicle emissions by raising automotive fuel prices and thus encouraging motorists to drive less and at slower speeds. In doing so, it would also address other social costs associated with driving, including those from other polluting emissions, accidents, noise, and congestion. The CAFE standards would have the opposite effect—they would encourage driving by reducing fuel costs. However, that "rebound" effect may be small.[4]

[1] Section 102 of the Energy Independence and Security Act of 2007, 49 U.S.C. § 32902(b)(2)(A), 121 Stat. 1499.

[2] See Congressional Budget Office, *The Economic Costs of Fuel Economy Standards Versus a Gasoline Tax* (December 2003).

[3] Because the new CAFE standards keep the distinction between cars and light trucks (automakers must meet each type of standard separately), under some circumstances a CO_2 price could affect the average fuel economy of new vehicles even with relatively stringent standards in place. If a high CO_2 price caused enough consumers seeking better fuel economy to switch from buying a new truck to buying a new car, the combined average fuel economy would go up.

[4] See Kenneth A. Small and Kurt Van Dender, "Fuel Efficiency and Motor Vehicle Travel: The Declining Rebound Effect," *Energy Journal*, vol. 28, no. 1 (2007), pp. 25–51.

That comparison suggests that gasoline prices might have to rise above \$6.50 per gallon—for example, from a CO_2 price that added \$2.00 or \$2.50 per gallon to gasoline prices—for the average fuel economy of new vehicles in the United States to approach the 35 mpg that the new CAFE standards will require. But the CO_2 prices contemplated in current U.S. climate legislation and in prominent international policy analyses would add much less than \$2.00 to the price of gasoline. Thus, such pricing, by itself, would probably not increase average fuel economy beyond what the CAFE standards will require.

CBO has estimated that under S. 2191, the America's Climate Security Act of 2007, the price of a CO2 emissions permit would rise from about \$23 per metric ton in 2009 to about $44 in 2018 as the stringency of the bill's cap on greenhouse-gas emissions was gradually increased.[23] Such permit prices would raise gasoline prices by about 20 cents per gallon in 2009 and 40 cents per gallon in 2018.[24] A recent report by the Intergovern-mental Panel on Climate Change (IPCC) suggests that a permit price of as much as $80 per ton of CO_2 might be necessary by 2030 to reduce emissions enough to achieve a stabilized climate by 2100. That pricing policy would add about 70 cents per gallon to the price of gasoline in 2030.[25] Even the much greater and much earlier reductions called for in the *Stern Review on the Economics of Climate Change* (requiring a current estimated permit price of $95 per ton of CO_2, rising to $191 per ton by 2050 and higher after that) would not cause gasoline prices in the United States to be as high as they already are in Europe.[26] The permit prices in the Stern report would add roughly $0.85 to $1.70 per gallon to gasoline prices over the next four decades.

The rising demand for more fuel-efficient vehicles is a trend that is likely to be reinforced as automakers gradually redesign more of their vehicles to better satisfy that demand. Over time, they will offer improved fuel economy for a wider array of vehicles appealing to a broader spectrum of consumer tastes.[27] Although a CO_2 emissions price would have relatively little effect on vehicle emissions, it would stimulate additional research and development of technologies for improving fuel efficiency, an effect that the new fuel economy standards will also have. New fuel-efficiency technologies will, in turn, create additional opportunities for reducing atmospheric CO_2 concentrations and stabilizing Earth's climate.

End Notes

[1]Congressional Budget Office, *Effects of Gasoline Prices on Driving Behavior and Vehicle Markets* (January 2008).

[2] CAFE standards specify the minimum average level of fuel economy that each automaker must achieve for the passenger vehicles it sells in the United States in a given model year.

[3] Energy Information Administration, http://tonto.eia.doe.gov/ dnav/pet/pet_pri_gnd_a_ epmr_pte_cpgal_w.htm .

[4] The Department of Transportation estimates that for the first quarter of 2008, U.S. motorists used about 1.3 percent less gasoline and 7 percent less diesel fuel than during the same period in 2007. See www.fhwa.dot.gov/pressroom/fhwa0817.htm .

[5] Department of Transportation, www.fhwa.dot.gov/ohim/tvtw/ 08juntvt/08juntvt.pdf.

[6] Estimates range from 0.3 percent to 0.8 percent, averaging about 0.6 percent. See Jonathan E. Hughes, Christopher R. Knittel, and Daniel Sperling, *Evidence of a Shift in the Short-Run Price Elasticity of Gasoline Demand*, Research Report UCD-ITS-RR-06-16 (University of California, Davis, Institute of Transportation Studies, 2006); and Kenneth A. Small and

Kurt Van Dender, "Fuel Efficiency and Motor Vehicle Travel: The Declining Rebound Effect," *Energy Journal*, vol. 28, no. 1 (2007), pp. 25–51.

[7] For faster traffic, CBO examined 85th percentile speeds because, according to the Federal Highway Administration, all states and most local agencies use the 85th percentile as a primary criterion in establishing their speed limits. For symmetry, CBO used 15th percentile speeds for slower traffic.

[8] Since the beginning of 2008, several major trucking companies have announced that, as a matter of company policy, they would reduce their trucks' top-end cruising speeds by about 3 mph, on average, by adjusting the trucks' speed-governing devices.

[9] See Department of Energy, P*olicies and Measures for Reducing Energy Related Greenhouse Gas Emissions: Lessons from Recent Literature*, DOE/PO-0047 (July 1996); and Small and Van Dender, "Fuel Efficiency and Motor Vehicle Travel."

[10] See Environmental Protection Agency, *Light-Duty Automotive Technology and Fuel Economy Trends: 1975 Through 2007*, EPA420-R-08-015 (September 2008), Table 1, p. 9, www.epa.gov/ oms/fetrends.htm .

[11] See National Highway Traffic Safety Administration, www.nhtsa.dot.gov.

[12] MSRPs may differ from the actual purchase prices negotiated between consumers and automobile dealers, including rebates and incentives. CBO did not have such data for this analysis, however.

[13] See Congressional Budget Office, *Effects of Gasoline Prices on Driv-ing Behavior and Vehicle Markets*, Box 2-2, pp. 21–22. The prices of used vehicles have been changing in the same way, and for some larger vehicles the drop in price has been particularly dramatic. For example, the wholesale prices of some 2005 model-year SUVs and pickup trucks fell by more than 20 percent over the first half of 2008, according to *Automotive News* (June 23, 2008).

[14] That includes a 7 percent reduction in emissions by entities subject to S. 2191, America's Climate Security Act of 2007 (see CBO's April 10, 2008, cost estimate at Error! Hyperlink reference not valid.Error! Hyperlink reference not valid. , plus additional and proportionately greater reductions from other, lower-cost sources—primarily via carbon sequestration and reduced emissions from landfills.

[15] CBO calculated the increase of 25 cents in the price of gasoline on the basis of about 20 pounds of CO_2 released per gallon of gasoline consumed. In theory, consumers and producers of gasoline would share that cost. But because gasoline consumption is relatively unresponsive to price in the short run, in practice consumers of gasoline would pay almost all of the CO_2 price.

[16] On the basis of the recent demand-response estimates cited in note 6, a 6 percent increase in gasoline prices would reduce consumption by only around 0.4 percent in the short run.

[17] A CO_2 price would have a much greater effect on the price of coal (a primary fuel in electricity generation) than on the price of gaso-line, simply because a dollar's worth of coal contains more carbon than a dollar's worth of gasoline. Moreover, the marginal costs of reducing CO_2 emissions may also be lower for coal-powered electricity generation and other sources than for vehicles.

[18] See CBO (January 2008), pp. xxi. Some analysts believe that con-sumers underestimate the value of fuel savings from improved fuel economy. If so, pricing CO_2 would be less than ideally effective against vehicle emissions—and a higher CO_2 price would be required to achieve a given reduction in vehicle emissions than if consumers valued fuel savings correctly.

[19] In terms of the European Union's voluntary de facto fuel-consumption standard, the 2006 average fuel economy in Europe was 160 grams of CO_2 per kilometer (km)—about 7 liters of gaso-line, or 6 liters of diesel fuel, per 100 km. The CO_2 rate converts to about 34 mpg. The 38-mpg value—an estimate of what the European average would be if measured using the U.S. test cycle—is used for comparison with the United States. For fuel economy averages, test-cycle differences, and conversion factors, see International Council on Clean

Transportation, *Passenger Vehicle Greenhouse Gas and Fuel Economy Standards: A Global Update* (July 2007), www.lowcvp.org.uk/assetsICCT_ GlobalStandards_2007.pdf.

[20] In 1998, when average fuel economy in Europe was about 180 g CO_2 /km, the European Union adopted a voluntary standard of 140 g CO_2 by 2008, which has not been achieved. It is now debating the adoption of either a mandatory fuel economy standard of 130 g CO_2 by 2012 or of 125 g CO_2 by 2015.

[21] European gasoline taxes are averages for 2002 and 2008 among the 15 countries that joined the European Union before 2004. See, respectively, *Fuel Taxation* (August 17, 2004; updated November 6, 2006), www.euractiv.com/en/taxationarticle-117495 ; and European Commission, *Excise Duty Tables* (July 2008), ec.europa.eu/taxation_ customs/resources documents/taxation/excise_duties_duties-part_II_energy_products-en.pdf. U.S. gasoline taxes vary by state. For the average U.S. fuel tax (as of January 2008), see American Petroleum Institute, www.api.org/policy/tax/stateexcise/ upload

[22] See Lee Schipper, *Automobile Fuel-Economy and CO2 Emissions in Industrialized Countries: Troubling Trends Through 2005/6*, World Resources Institute (2008), pdf.wri.org/ automobile-fuel-economy-co2-industrialized-countries.pdf. For environmental reasons, automakers have not been able to sell many diesel-powered vehicles in the United States, although that is set to change with the development of low-sulfur diesel fuels that can satisfy the more stringent U.S. standards for particulate emissions.

[23] See CBO's cost estimate for S. 2191, America's Climate Security Act of 2007 (April 10, 2008), www.cbo.gov/ftpdocs/91xx/ doc9120/s2191.pdf.

[24] EPA has also analyzed S. 2191 and estimates CO_2 prices that would add about $0.53 per gallon in 2030 and $1.40 per gallon in 2050. See www.epa.gov/climatechange/downloads/ s2191_EPA_Analysis.pdf.

[25] See International Panel on Climate Change, Climate Change 2007: Synthesis Report, p. 59, www.ipcc.ch/pdf/assessmentar4/syr/ar4_syr.pdf.

[26] See www.hm-treasury.gov.uk/independent_reviews/ stern_review_ economicsstern_ review_ report.cfm.

[27] For example, hybrid technology is being introduced on larger vehicles, including the Chevrolet Tahoe and Cadillac Escalade— big SUVs with combined fuel economy ratings of 21 mpg and 20 mpg, respectively.

In: Policy Option Issues for CO2 Emissions ISBN: 978-1-60741-381-3
Editors: Nikolaus Vogler

Chapter 2

ISSUES IN DESIGNING A CAP-AND-TRADE PROGRAM FOR CARBON DIOXIDE EMISSIONS

Peter R. Orszag

Chairman Rangel, Congressman McCrery, and Members of the Committee, thank you for the invitation to testify this morning on reducing the economic costs involved in addressing climate change. If policymakers adopt a cap-and-trade program to reduce carbon dioxide (CO_2) and other greenhouse gases, the economic costs will depend on several specific design features of the program.

Global climate change is one of the nation's most significant long-term policy challenges. Human activities are producing increasingly large quantities of greenhouse gases, particularly CO_2. The accumulation of those gases in the atmosphere is expected to have potentially serious and costly effects on regional climates throughout the world. Although the magnitude of such damage remains highly uncertain, there is growing recognition that some degree of risk exists for the damage to be large and perhaps even catastrophic.

The risk of potentially catastrophic damage from climate change can justify taking action to reduce that risk in much the same way that the hazards we all face as individuals motivate us to buy insurance. Some of society's resources may best be devoted to addressing climate change even if the most severe damage ultimately does not materialize.

Reducing greenhouse-gas emissions would be beneficial in limiting the degree of risk associated with climate change, especially the risk of significant damage. However, decreasing those emissions would also impose costs on the economy—in the case of CO_2, because much economic activity is based on fossil fuels, which release carbon in the form of that gas when they are burned. Much of those costs will be passed along to consumers in the form of higher prices for energy and energy-intensive goods.

Designing a cap-and-trade program to achieve such reductions would include important decisions about whether to sell or give away allowances. Those rights to emit greenhouse gases would have substantial value, and policymakers' choices about how to allocate them could have significant effects on the federal budget and on how the gains and losses brought about by the program were distributed among U.S. households. If policymakers chose to sell the allowances, they could use the revenue that would arise in many different ways, including to offset other taxes, to assist workers or low-income households that might be adversely affected by the cap, to support other legislative priorities, or to reduce the budget deficit. Policymakers would also need to decide whether to include provisions to help contain the cost of the policy by allowing firms flexibility as to when they reduced their emissions and whether to include provisions to address effects on international trade, particularly for energy-intensive goods. My testimony makes the following key points about those issues:

- A cap-and-trade program could raise significant revenue because the value of the allowances created under the program would probably be substantial. For example, the Congressional Budget Office (CBO) estimated that the value of the allowances under the cap-and-trade proposal that went to the Senate floor in June would be roughly \$112 billion once the cap took effect in 2012 and would increase as the cap became more stringent.
- Issuing allowances to entities at no charge, provided that the recipients can readily convert the allowances into cash, is economically equivalent to selling the allowances and dedicating the revenue to those same entities. That equivalency is likely to hold when the allowances can be sold in a large and liquid secondary market, as would be the case under most cap-and-trade programs for greenhouse gases. CBO's estimate of the federal cost of two recent bills considered by the Senate, S. 2191 and S. 3036, reflected that

equivalency by recording the value of most of the freely allocated allowances as both revenues and outlays.

- Policymakers' decisions about how to allocate the allowances could have significant effects on the overall economic cost of capping CO_2 emissions and on the distribution of gains and losses among U.S. households. Giving away a large share of the allowances to companies that produce fossil fuels or energy-intensive goods could be more costly to the economy and more regressive than selling them. That approach would preclude using the value of the allowances to create additional incentives for economic activity. It could also create "windfall profits" for shareholders, while not preventing the cap from causing price increases that would disproportionately affect low-income people.
- If the government chose to sell emission allowances, it could use some of the revenue to offset the disproportionate economic burden that higher prices would impose on low-income households. Selling allowances could also significantly lessen the overall economic impact of a CO_2 cap. Evidence suggests that the economic cost of a 15 percent cut in U.S. emissions (not counting any benefits from mitigating climate change) might be more than twice as large if policymakers gave the allowances away than if they sold the allowances and used the revenue to lower current taxes on labor or capital that discourage economic activity, such as income or payroll taxes. Likewise, dedicating the allowance revenue to reduce the federal deficit could lower the overall economic cost.
- Policymakers could help reduce the cost of achieving any given long-term target for reducing emissions if they included provisions in a cap-and-trade program that allowed firms some degree of flexibility about when emission reductions take place. Such provisions would augment the flexibility about where and how emission cuts are made that is intrinsic to a cap-and-trade program. Timing flexibility would allow firms to reduce emissions more when the cost of doing so was low and would provide firms leeway to reduce their efforts when costs were high. One method of providing timing flexibility would be to set a ceiling and a floor for allowance prices. The ceiling would limit firms' expenses when the cost of cutting emissions was high, and the floor would automatically tighten the cap (and thereby increase emission reductions) when the cost of cutting emissions was low. Policymakers could periodically adjust the speed at which the price

ceiling and floor increased to ensure that emission reductions were on track for achieving a long-term target.

- Energy-intensive U.S. industries that face foreign competition (for example, the steel and aluminum industries) could lose sales to imports from countries that did not have similarly stringent policies to reduce greenhouse gases. That substitution of imports for U.S. production could reduce the environmental benefits of the policy, because it would result in emission increases from countries with less stringent policies. Some proposals would address those concerns by providing transitional assistance to manufacturers of energy-intensive products in the United States or requiring importers of those products to purchase allowances. Those proposals could, in the short run, protect domestic manufacturers from being disproportionately harmed and limit the loss of intended environmental benefits. Even so, questions remain about whether the proposals could be effectively implemented in a way that would be consistent with U.S. obligations under its agreements with the World Trade Organization.

How a Cap-and-Trade Program Would Work

Under a cap-and-trade program, policymakers would set a limit on total emissions during some period and would require regulated firms to hold rights, or allowances, to the emissions permitted under that cap. (Each allowance would entitle companies to emit one ton of CO_2 or to have one ton of carbon in the fuel that they sold.) After the allowances for a given period were distributed, firms would be free to buy and sell the allowances among themselves. Firms that were able to reduce emissions most cheaply would profit from selling allowances to firms that had relatively high abatement costs. The trading aspect of the program would lead to substantial cost savings relative to command-and-control approaches—which would mandate how much entities could emit or what technologies they should use—because it would provide more flexibility about where and how emission reductions were achieved.

A cap-and-trade program has been implemented at the federal level in the United States to limit emissions of sulfur dioxide (which contribute to acid rain). That program has been in effect since 1995 and is widely perceived to have reduced emissions at a significantly lower cost than would have been the case if lawmakers had chosen to rely on a command-and-control approach. Several states have considered, or adopted, plans for a cap-and-trade program for CO_2 emissions, but none is yet operational. A cap-and-trade program for CO_2 emissions is currently in operation in the European Union as part of its effort to comply with emission limits under the initial phase of the Kyoto Protocol, which spans 2008 to 2012.

The Potential Value and Budgetary Treatment of Allowances

In establishing a cap-and-trade program, policymakers would create a new commodity: the right to emit CO_2. The emission allowances would have substantial value. On the basis of a review of the existing literature and the range of CO_2 policies now being debated, CBO estimated that by 2020, the value of those allowances could total between $50 billion and $300 billion annually (in 2006 dollars). The actual value would depend on various factors, including the stringency of the cap, the possibility of offsetting CO_2 emissions through carbon sequestration or international trading of allowances, and other features of the specific policy that was selected.[1] On June 2, 2008, CBO estimated that the value of the allowances created under S. 3036 would be roughly $112 billion once the proposed program took effect in 2012; in subsequent years, the aggregate value of the allowances would be even greater.

Policymakers would need to decide how to allocate the allowances that corresponded to each year's CO_2 cap. One option would be to have the government capture their value by selling the allowances, as it does with licenses to use the electromagnetic spectrum. Another possibility would be to give the allowances to energy producers or some energy users at no charge. The European Union has used that second approach in its 2-year-old cap-and-trade program for CO_2 emissions, and nearly all of the allowances issued under the 13-year-old U.S. cap-and-trade program for sulfur dioxide emissions are distributed in that way.

The budgetary treatment of allowances that are auctioned is straightforward: The auctions generate receipts for the federal government, and those amounts are recorded as revenues. In some cases, allowances that are given away by the government should, in CB O's view, also be reflected in the federal budget (recorded both as revenues and outlays). That treatment is appropriate when the allowance recipients would be able to immediately and easily convert the allowances into cash by selling them in a large and liquid secondary market. In such cases, distributing allowances at no charge to specific firms or individuals is, in effect, equivalent to auctioning the allowances and then distributing the auction proceeds to those firms or individuals. Treating allowances issued at no charge as both revenues and outlays reflects the equivalency of those two scenarios. CBO applied that budgetary treatment to most of the allowances freely allocated under S. 2191 and S. 3036. (In contrast, the proceeds associated with the allowances allocated free of charge to producers and importers under smaller, more constrained cap-and-trade programs—such as the cap-and-trade program for hydro- fluorocarbons proposed under S. 2191 and S. 3036—should not be recorded in the budget, CBO believes, primarily because the market created for such allowances would be relatively illiquid and, therefore, the allowances would be less like cash.)

The Distributional Consequences of a Cap-and-Trade Program

Whether policymakers decided to sell the allowances or give them away would have significant implications for the distribution of gains and losses among U.S. households. The ultimate distributional impact of a cap-and-trade program would be the net effect of two distinct components: the distribution of the costs of the program (including the cost of paying for the allowances) and the distribution of the allowances' value. Market forces would determine who bore the costs of a cap-and-trade program, but policymakers would determine who received the value of the allow-ances. The ultimate effect could be either progressive or regressive, imposing disproportionately large burdens on high-income or low-income households, respectively.

Market Forces Would Determine Who Bore the Costs of a Cap

Obtaining allowances—or taking steps to cut emissions to avoid the need for such allowances—would become a cost of doing business for firms that were subject to the CO_2 cap. However, those firms would not ultimately bear most of the costs of the allowances. Instead, they would pass them along to their customers (and their customers' customers) in the form of higher prices.

By attaching a cost to CO_2 emissions, a cap-and-trade program would thus lead to price increases for energy and energy- intensive goods and services, the production of which contributes the most to those emissions. Such price increases would stem from the restriction on emissions and would occur regardless of whether the government sold emission allowances or gave them away. Indeed, the price increases would be essential to the success of a cap-andtrade program because they would be the most important mechanism through which businesses and households would be encouraged to make investments and behavioral changes that reduced CO_2 emissions.

The rise in prices for energy and energy-intensive goods and services would impose a larger burden, relative to income, on low-income households than on high-income households.

Table 1. Effects on U.S. Households of the Higher Prices Resulting from a 15 Percent Cut in CO_2 Emissions

	Average for Income Quintile				
	Lowest	Second	Middle	Fourth	Highest
Annual Cost Increase in 2006 Dollars	680	880	1,160	1,500	2,180
Annual Cost Increase as a Percentage of Income[a]	3.3	2.9	2.8	2.7	1.7

Source: Congressional Budget Office, *Who Gains and Who Pays Under Carbon-Allowance Trading? The Distributional Effects of Alternative Policy Designs* (June 2000).

Notes: These numbers do not reflect any of the benefits from reducing climate change.

The policy examined here is a cap-and-trade program designed to lower U.S. carbon dioxide (CO_2) emissions by 15 percent from 1998 levels. (CBO performed the analysis in 2000 and used 1998 emission levels so that the distributional effects could be based on actual, rather than projected, data on consumer spending and taxes.) CBO assumed that the full cost of cutting emissions would be passed along to consumers in the form of higher prices and that the price increase for a given product would be proportional to the amount of CO_2 emitted from the fossil fuels used in its production.

These numbers reflect data on each quintile's cash consumption and estimates of cash income. (A quintile contains one-fifth of U.S. households arrayed by income.) Because of data limitations, the numbers should be viewed as illustrative and broadly supportive of the conclusions in this analysis rather than as precise estimates.

a. The cost increases are equivalent to percentage declines in households' after-tax income.

For example, without incorporating any benefits to households from lessening climate change, CBO estimated that the price increases resulting from a 15 percent cut in CO_2 emissions would cost the average household in the lowest one-fifth (quintile) of all households arrayed by income slightly more than 3 percent of its income; such increases would cost the average household in the top quintile just under 2 percent of its income (see Table 1).[2]

The higher prices that would result from a cap on CO_2 emissions would reduce demand for energy and energy-intensive goods and services and thus create losses for some current investors and workers in the sectors of the economy that supply such products. Investors might see the value of their stock decline, and workers could face the risk of unemployment as jobs in those sectors were cut. Stock losses would tend to be widely dispersed among investors, because shareholders typically diversify their portfolios. In contrast, the costs borne by workers would probably be concentrated among relatively few households and, by extension, their communities.

Policymakers Would Determine Who Received the Value of the Allowances

Although the price increases triggered by a cap-and-trade program for CO_2 emissions would be regressive, the program's ultimate distributional effect would depend on policymakers' decisions about how to allocate the allowances. As noted above, those allowances would be worth tens or hundreds of billions of dollars per year. Who received that value would depend on how the allowances were distributed.

Lawmakers could more than offset the price increases experienced by low-income households or the costs imposed on workers in particular industrial sectors by providing for the sale of some or all of the allowances and using the revenue to pay compensation. From analyzing the ultimate distributional effects of a cap-and-trade program that would reduce CO_2 emissions in the United States by 15 percent, CBO concluded that lower-income households could be better off (even without counting any benefits from reducing climate change) as a result of the policy if the government chose to sell the allowances and use the revenue to pay an equal lump-sum rebate to every household in the United States. In that case, the size of the rebate would be larger than the average increase in low-income households' spending on energy and energy-intensive goods.[3] Such a strategy would increase average income for households in the lowest income quintile by about

2 percent (see the top panel of Figure 1). At the same time, average income for households in the top quintile would fall by less than 1 percent, CBO estimates.

In contrast, if lawmakers chose to use the allowances to decrease corporate income taxes, the effect would be significantly more regressive than the initial price increases. Because low-income households pay relatively little in corporate taxes, the reduction in corporate tax rates would not offset their increased spending on energy and energy- intensive goods. Households in the top income quintile, however, would experience an increase in after-tax income as a result of the policy. Should policymakers decide to use the revenue from selling allowances to decrease payroll taxes, the effect (not shown in the figure) would be regressive as well, although less so than for a cut in corporate taxes.[4]

Giving all or most of the allowances to energy producers to offset the potential losses of investors in those industries—as was done in the cap-and-trade program for sulfur dioxide emissions—would also exacerbate the regressivity of the price increases. On average, the value of the CO_2 allowances that producers would receive would more than compensate them for any decline in profits caused by a drop in demand for energy and energy-intensive goods and services whose production causes emissions. As a result, the companies that received allowances could experience windfall profits.

For example, in 2000, CBO estimated that if emissions were reduced by 15 percent, as in the scenario discussed above, and all of the allowances were distributed free of charge to producers in the oil, natural gas, and coal sectors, the value of the allowances would be 10 times as large as coal, oil, and natural gas producers' combined profits in 1998. Profits for those industries have climbed substantially since then, yet the value of the allowances associated with the policy that CBO analyzed would still be large relative to those producers' profits.[5] Because the additional profits from the allowances' value would not depend on how much a company produced, such profits would be unlikely to prevent the declines in production and resulting job losses that the price increases (and resulting drop in demand) would engender.

In addition, those profits would accrue to shareholders, who are primarily from higher-income households, and would more than offset those households' increased spending on energy and energy-intensive goods and services. Low-income households, by contrast, would benefit little if allowances were given to energy producers for free, and they would still bear a disproportionate burden from the price increases that would nonetheless occur. Thus, giving away allowances would be significantly regressive, making higher-income households better off as a result of the cap-andtrade policy while making lower-income households worse off.

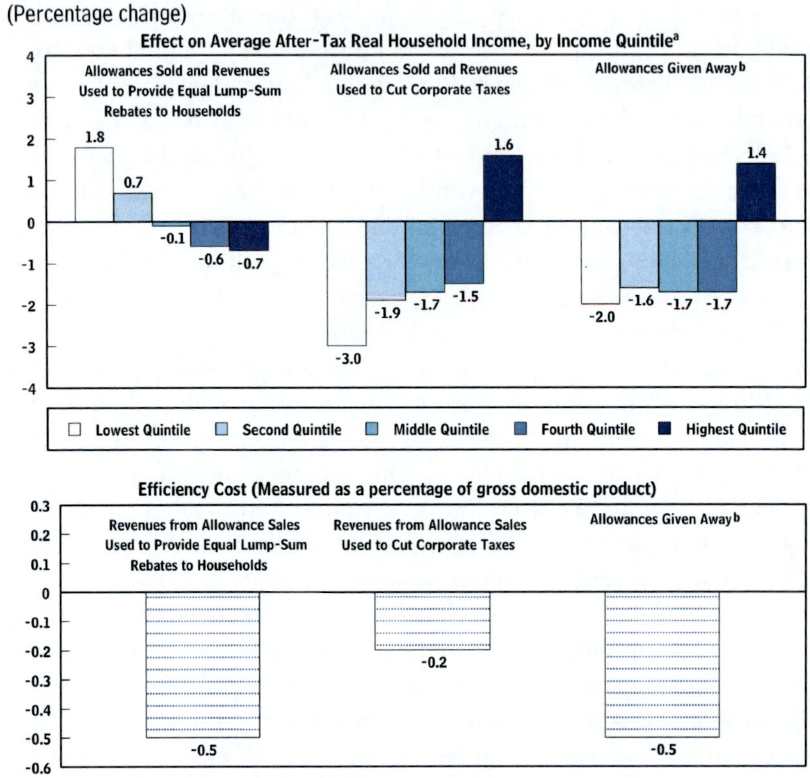

Sources: Congressional Budget Office (top panel); Terry M. Dinan and Diane Lim Rogers (bottom panel), "Distributional Effects of Carbon Allowance Trading: How Government Decisions Determine Winners and Losers," *National Tax Journal*, vol. 55, no. 2 (June 2002).

Notes: These figures do not reflect any of the benefits from reducing climate change.

The policy examined here is a cap-and-trade program designed to reduce carbon dioxide (CO_2) emissions by 15 percent from 1998 levels. (CBO performed the analysis in 2000 and used 1998 emission levels so the distributional effects could be based on actual, rather than projected, data on consumer spending and taxes.) In the top panel, the costs of the cap-and-trade policy are shown as decreases in real household income, measured as a percentage of after-tax income before the policy change. Those numbers reflect data on each quintile's cash consumption and estimates of cash income. (A quintile contains one-fifth of U.S. households arrayed by income.) Because of data limitations, those numbers should be viewed as illustrative and broadly supportive of the conclusions in this analysis rather than as precise estimates.

a. Indicates the net effect of households' increased expenditures because of cap-induced price increases and the income that households would receive as a result of the allowance-allocation strategy.

b. These estimates assume that the government would use any positive net revenue remaining after accounting for ways in which the policy affected the federal budget to provide equal lump-sum rebates to households. The results would be more regressive if the government used any positive net revenue to decrease corporate taxes or payroll taxes.

Figure 1. Effects of a 15 Percent Cut in CO_2 Emissions, with the Allowances' Value Used in Various Ways.

REDUCING THE OVERALL ECONOMIC IMPACT OF A CO_2 CAP

Restricting CO_2 emissions would impose costs on the economy. Lawmakers could help minimize those costs by using the allowances' value in ways that would benefit the economy and by allowing firms some degree of flexibility about when emission reductions must be made.

Using the Allowance Value to Reduce the Total Economic Cost

The ways in which lawmakers allocate the revenue from selling emission allowances would affect not only the distributional consequences of a cap-and-trade policy but also its total economic cost. For instance, the government could use the revenue from auctioning allowances to reduce existing taxes that tend to dampen economic activity—primarily, taxes on labor, capital, or personal income. As research indicates, a CO_2 cap would exacerbate the economic effects of such taxes: The higher prices caused by the cap would lower real (inflation-adjusted) wages and real returns on capital, which would be equivalent to raising marginal tax rates on those sources of income. Using the value of the allowances to reduce such taxes could help mitigate that adverse effect of the cap. Alternatively, policymakers could choose to use the revenue from auctioning allowances to reduce the federal deficit. If that reduction lessened the need for future tax increases, the end result could be similar to dedicating the revenue to cuts in existing taxes.

The decision about whether or not to sell the allowances and use the proceeds in ways that would benefit the economy could have a significant impact. For example, researchers have estimated that the efficiency cost (discussed below) of a 15 percent cut in emissions could be reduced by more than half if the government sold the allowances and used the revenue to lower corporate income taxes, rather than devoting the revenue to providing lump-sum rebates to households or giving the allowances away (see the bottom panel of Figure 1). The efficiency cost of a policy reflects the economic losses that occur because prices in the economy are distorted in that they do not reflect the (nonenvironmental) resources used in their production. That cost includes decreases in the productive use of labor and capital as well as costs (both monetary and nonmonetary) associated with reducing emissions. To provide perspective on the magnitude of such efficiency costs, they are depicted as a share of gross domestic product.

Allowing Flexibility in the Timing of Emission Reductions to Lower Costs

In its most inflexible form, a cap-and-trade program would require that a specified cap on emissions was met each year. That lack of flexibility would increase the cost of achieving any long-term goal because it would prevent firms from responding to year- to-year differences in conditions that affected costs for reducing emissions, such as fluctuations in economic activity, energy markets, the weather (for example, an exceptionally cold winter would increase the demand for energy and make meeting a cap more expensive), and the technologies available for reducing emissions.

In contrast, the key issue from an environmental perspective involves the emissions and concentrations of greenhouse gases over the long term, not the year-to-year fluctuations in emissions. In other words, limiting global climate change will entail substantially reducing the amount of greenhouse gases that accumulate in the atmosphere over the next several decades, but the benefits of doing so are largely independent of the annual pattern of those reductions.[6] Consequently, a cap-and-trade program could achieve roughly the same level of benefits at a significantly lower cost if it provided firms with an incentive to make greater reductions in emissions at times when the cost of doing so was low and allow them leeway to lessen their efforts when the cost was high.

Including features in a cap-and-trade program that enabled firms to reduce emissions less when the cost was high and more when the cost was low could also reduce the volatility of allowance prices. Experience with cap-and-trade programs has shown that price volatility can be significant. For example, one researcher found that the price of sulfur dioxide allowances under the U.S. Acid Rain Program was significantly more volatile than stock prices between 1995 and 2006 (see Figure 2).[7]

Price volatility could be particularly problematic with CO_2 allowances because fossil fuels play such an important role in the U.S. economy. In 2006, fossil fuels accounted for 85 percent of the energy consumed in the United States. CO_2 allowance prices could affect energy prices, inflation rates, and the value of imports and exports. If those prices were volatile, they could have disruptive effects on markets for energy and energy-intensive goods and services and could make investment planning difficult.

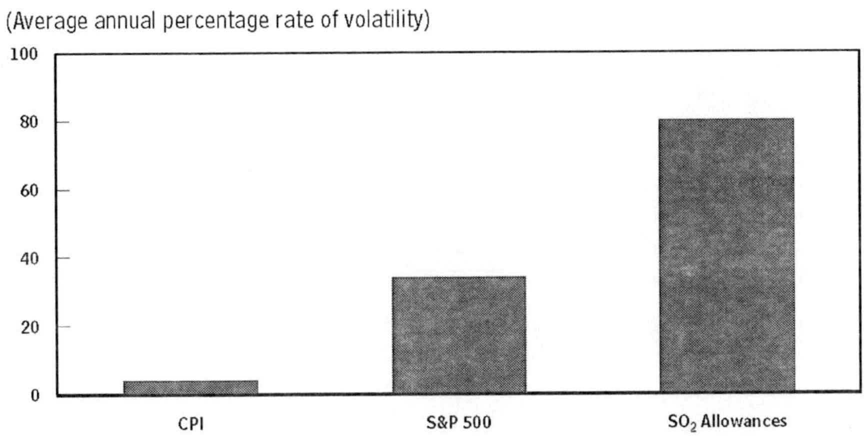

Source: Congressional Budget Office based on William D. Nordhaus, "To Tax or Not to Tax: Alternative Approaches to Slowing Global Warming," *Review of Environmental Economics and Policy*, vol. 1, no. 1 (Winter 2007), pp. 26–44.

Note: Volatility is calculated as the annualized absolute logarithmic month-to-month change in the consumer price index (CPI), the stock price index for the Standard & Poor's 500 (S&P 500), and the price of sulfur dioxide (SO_2) allowances under the U.S. Acid Rain Program.

Figure 2. Volatility of SO_2 Allowance Prices and Selected Other Prices, 1995 to 2006.

Design Features Providing Flexibility in the Timing of Emission Reductions

Recent proposals for cap-and-trade proposals include a variety of design features that would provide firms or regulators with flexibility in the timing of emission reductions, thereby reducing the economic costs of limiting greenhouse-gas emissions.

A Price Ceiling and a Price Floor

A combined price ceiling and price floor offers one method of allowing timing flexibility and thereby reducing the economic burden of achieving any desired target for cumulative emissions:

- Setting a ceiling, or safety valve, for the price of allowances could prevent the cost of reducing emissions from exceeding either the best available estimate of the environmental benefits or the cost that policymakers considered acceptable. The government could maintain a price ceiling by selling companies as many allowances as they would like to buy at the safety-valve price.
- Similarly, policymakers could prevent the price of allowances from falling too low by setting a price floor. If the government chose to auction a significant share of the allowances, it could specify a so-called reserve price and withhold allowances from the auction as needed to maintain that price. The efficiency advantage of a price floor would stem from the fact that it could prevent the cost of emission reductions from falling below the expected benefits or below the level of effort that policymakers intended.

A cap-and-trade program that included both a ceiling and a floor for allowance prices could achieve a long-term target for emissions while minimizing both the overall cost of achieving the target and price volatility. Under such a program, policymakers would specify annual emission targets as well as a ceiling and a floor for the price of allowances for each year. Regulators could adjust the levels of the price ceiling and floor periodically (for example, every five years) to ensure that emission reductions were on track for achieving the long-run target. For example, the rate at which the price floor or ceiling rose over time could be increased if regulators determined that the reductions in the previous five-year period were significantly lower than the amount needed to achieve the long-term target. Alternatively, policymakers could include provisions in a cap-and-trade program that would automatically trigger adjustments in the price ceiling and floor. For example, the rate at which the price ceiling and floor rose could be based on the percentage gap between anticipated and actual emissions in the previous five-year period.

Figures 3 and 4 illustrate the effects of price ceilings and floors. The figures present a simple example of an inflexible cap each year relative to a system involving price ceilings and floors. In Figure 3, the results illustrate what happens if the cost of reducing emissions by 15 percent is twice as high or 50 percent lower than expected in any given year. Under an inflexible cap, the emission reductions are unaffected. Under a price ceiling, fewer emission reductions are undertaken when the cost is high; the result is lower economic costs that year but also less of a reduction in emissions. Under a price floor, more emission reductions are undertaken when the cost is low.

Figure 4 shows the results after one high-cost year and one low-cost year. Cumulative emission reductions are the same under the inflexible cap and the combined price ceiling and floor, but costs are substantially lower under the latter approach. The reason, again, is that more of the emission reductions are undertaken in the low-cost year under that approach.

Borrowing and Banking Allowances

An alternative but generally somewhat less effective approach to reducing economic costs involves allowing companies to borrow future allowances in high-cost years, thereby deferring emission reductions to later years. Borrowing allowances from future years would tend to reduce allowance prices in the current year but then raise prices in the future (because borrowing would allow smaller reductions now but require greater reductions later). Firms would want to borrow allowances only if they expected the price of allowances in the future to be sufficiently below the current price as to make deferring reductions profitable. Most proposals would impose limits on borrowing, furthermore, in part because of concerns about enforcement and questions about who would be liable if the firm that borrowed future allowances was unable to pay them back (if it declared bankruptcy, for example).

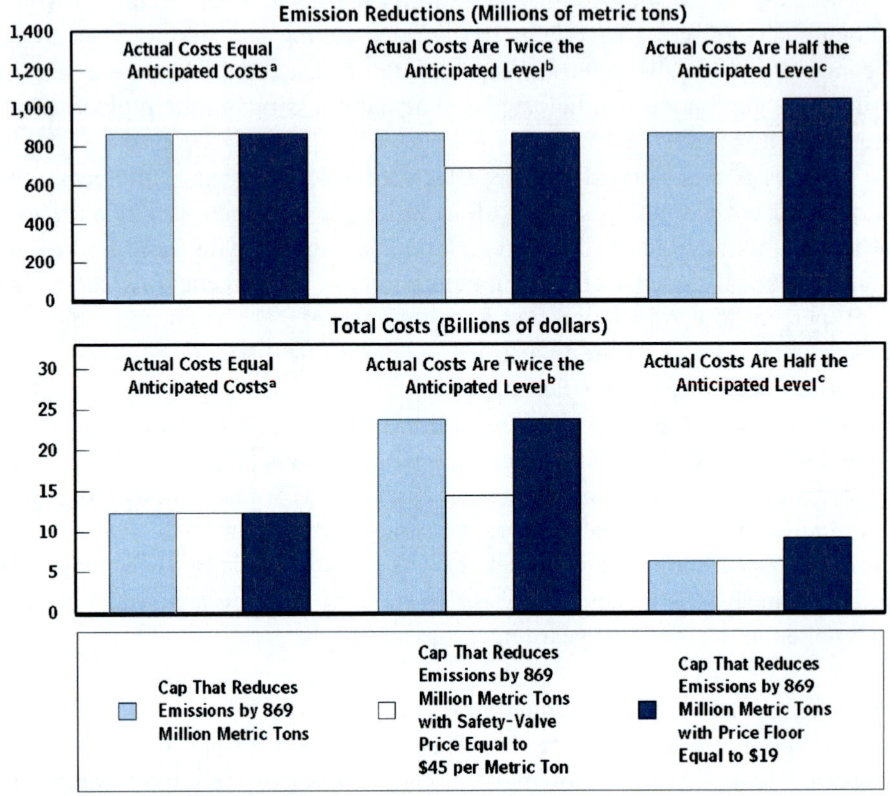

Source: Congressional Budget Office.

Notes: This example examines the emission reductions and total costs that would result in 2018, assuming that the policy covered only the United States. The cost of firms' emission reductions is derived from Mark Lasky, *The Economic Costs of Reducing Emissions of Greenhouse Gases: A Survey of Economic Models*, Congressional Budget Office Technical Paper No. 2003- 03 (May 2003).

A safety valve is a ceiling on the price of emission allowances.

a. Assumes that the actual marginal cost of reducing emissions by 869 million metric tons is $30 per metric ton, the cost that policymakers anticipated when they set the cap.

b. Assumes that the actual marginal cost of reducing emissions by 869 million tons is $60 per metric ton but that the safety valve induces less reductions (691 million tons instead of 869 million), up to a marginal cost of $45 per metric ton.

c. Assumes that the actual marginal cost of reducing emissions by 869 million tons is $15 per metric ton but that the price floor induces more reductions (1,047 million tons instead of 869 million) at a marginal cost of $19 per metricton.

Figure 3. Illustrative Comparison of Various Cap-and-Trade Policies to Reduce CO_2 Emissions by Roughly 15 Percent Under Different Cost Conditions in 2018.

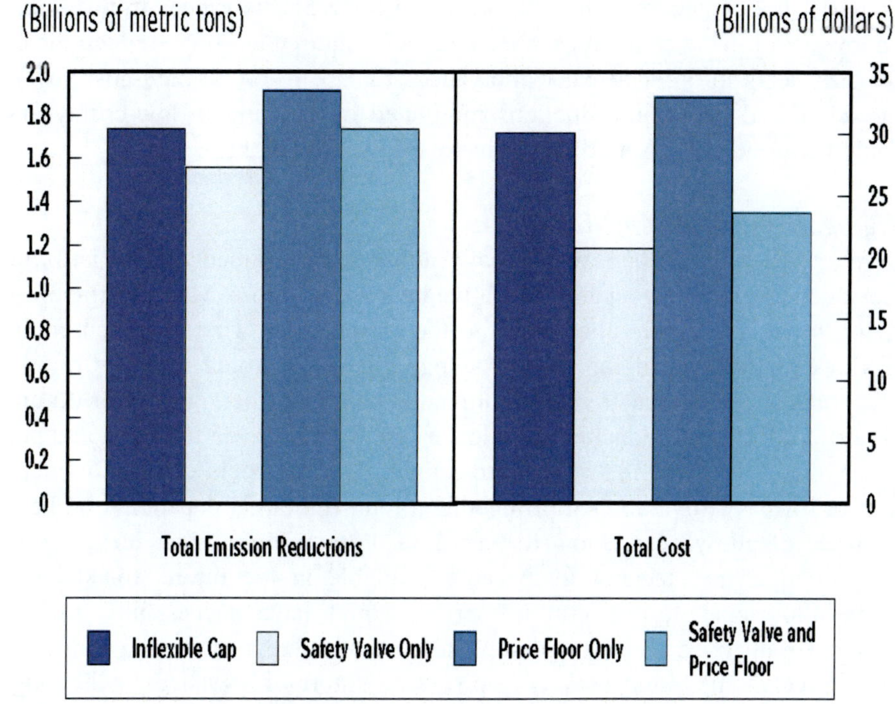

Source: Congressional Budget Office.

Notes: This example represents the cumulative emission reductions and costs over two years of a cap-and-trade policy that would reduce emissions of carbon dioxide by 869 million tons in each year (roughly a 15 percent reduction in 2018). The cost of firms' emission reductions is derived from Mark Lasky, *The Economic Costs of Reducing Emissions of Greenhouse Gases: A Survey of Economic Models*, Congressional Budget Office Technical Paper No. 2003-03 (May 2003).

A safety valve is a ceiling on the price of emission allowances.

For the high-cost year, CBO assumes that the marginal cost of reducing emissions by 869 million tons is $60 per metric ton but that the safety valve induces less reductions (691 million tons instead of 869 million), up to a marginal cost of $45 per metric ton.

For the low-cost year, CBO assumes that the marginal cost of reducing emissions by 869 million tons is $15 per metric ton but that the price floor induces more reductions (1,047 million tons instead of 869 million) at a marginal cost of $19 per metric ton.

Figure 4. Illustrative Comparison of Total Emission Reductions and Total Costs After One High-Cost and One Low-Cost Year

Similarly, policymakers could help keep the price of allowances from falling too low by allowing companies to exceed their required emission reductions in low-cost years in order to bank allowances for use in future high-cost years. The additional emission reductions motivated by banking in low-cost years would put upward pressure on the price of allowances in those years.

Aggregate Borrowing by Regulators

S. 2191 and S. 3036 would have addressed sustained high prices for allowances by allowing an administrative board, the Carbon Market Efficiency Board, to transfer future allowances to the current year. That approach could be viewed as a form of forced borrowing—that is, it would require firms to trade higher reductions in the future and, hence, higher prices for future allowances for lower reductions and lower prices today, even if they would not have voluntarily made this trade across time. Such transfers could ultimately raise or lower the overall cost of meeting a long-run target depending on how the price of allowances changed over time. For example, if an inexpensive low-carbon energy technology became available in the future, transferring future allowances to the current period would have successfully shifted emission reductions to a time when the cost of achieving them was lower. Alternatively, if policymakers borrowed future allowances with the expectation that such a technology would become available, but it did not, then the transfer could cause even more reductions to be made at a relatively high-cost time.

An alternative approach to the one embodied in those bills, which may be easier for regulators to implement efficiently, would be to adopt a system providing a combined price ceiling and price floor and to have the board be responsible for setting the ceiling and floor prices and for adjusting those price limits periodically as needed to achieve a long-term target.

Design Features Addressing Energy-Intensive Manufacturing Industries

Several bills introduced during the 110th Congress contain provisions that pertain to certain energy-intensive manufacturing industries that are subject to foreign competition. The proposals potentially address the concern that if a cap-and-trade system was adopted in the United States, manufacturing in certain industries could shift to countries with less stringent climate policies, which would undermine the environmental goals that the domestic policy was intended to achieve and harm U.S. manufacturers.

During 2006, about 30 percent of total U.S. greenhouse-gas emissions came from domestic manufacturing industries, either directly through the manufacturing process or indirectly through the use of electricity in manufacturing.[8] Some of those industries—such as those producing chemicals, glass, cement, iron and steel, aluminum, food, and paper and pulp, which accounted for about 15 percent of U.S. energy consumption—face significant foreign competition.

Reducing greenhouse-gas emissions would raise production costs for energy-intensive manufacturers by increasing the prices of the energy that they purchase or by causing them to invest in equipment to reduce their emissions. As a result, their sales, employment, and profits would probably decline, at least in the short run. Over the longer term, changes in investment and in the mix of output produced in the United States would mitigate those effects.

Unlike companies that face only limited foreign competition, such as electric utilities, manufacturers might have difficulty passing increased costs on to consumers because they face competitors in both the U.S. and international markets. If foreign competitors (either foreign-owned or subsidiaries of U.S. firms) did not experience similar increases in production costs because of less stringent emission standards abroad, U.S. output and employment in those industries would probably decline. Moreover, the environmental benefits of reducing U.S. emissions might be partially offset by increased emissions elsewhere, an effect known as "carbon leakage."

Several recent bills—including S. 1766, S. 3036, H.R. 6186, and H.R. 6316 (see Table 2)—have included two design features that are intended to cushion the economic impacts on energy-intensive manufacturing industries and to protect against carbon leakage: border adjustments and transitional assistance.

Border Adjustments

Recent bills have included provisions that, in effect, would require importers of certain goods from countries without climate policies to obtain allowances from an international reserve on the basis of the carbon emissions embodied in those goods. Such border adjustments would increase the cost of imports, raise their price in the U.S. market, and thus level the playing field for affected U.S. producers. In S. 1766 and S. 3036, the basis for calculating the number of reserve allowances required for imports was the extent to which carbon emissions in a given sector for a given country increased following the introduction of a climate policy in the United States. In more recent proposals, such as H.R. 6186 and H.R. 6316, the allowance requirement is based on the total emissions from a given sector in a given country. Some of the bills also would take into account the economic development status of the exporting country and the extent to which it had undertaken efforts to reduce greenhouse-gas emissions. In addition, many of the bills recently proposed would reduce the allowance requirement by the degree to which transitional assistance was given to domestic manufacturing.

By increasing the price of imports, border adjustments would limit any increase in CO_2 emissions from having unregulated imports displace goods produced in the United States. The requirements would give exporting countries the same incentive to reduce emissions in the production of their exports to the United States that the capand-trade program would give to U.S. producers.

Adjustments to exports are also possible but have not been included in recent proposals. Policymakers could provide allowances for free to U.S manufacturers to cover the emissions associated with the goods that they export. Such adjustments would lower compliance costs for U.S. exporters and could help mitigate any losses in U.S. exports that might otherwise occur as a result of a climate policy. When U.S. exports are already less carbon-intensive than foreign competitors' goods, export adjustments would serve to guard against carbon leakage in international markets.

When U.S. exports are more carbon-intensive than foreign competitors' goods, export adjustments could undermine the benefits from meeting a domestic cap on greenhouse gases. Export adjustments would also eliminate the incentive to reduce greenhouse-gas emissions from the manufacture of goods for export, making it more expensive to achieve a domestic cap.

The scope of coverage differs among the bills. They all cover energy-intensive primary goods (for example chemicals, glass products, cement products, iron and steel products, aluminum products, food products, and paper and pulp products), and some include various energy-intensive finished goods, such as vehicles. The wider the scope, the more difficult it becomes to calculate the number of reserve allowances required. Regulators would find it particularly challenging to estimate the carbon emissions embodied in finished goods, especially when inputs might come from different countries using different technologies.

Table 2. Comparison of Border Adjustments and Transitional Assistance in Recent Bills

Bill Number	Applicability[a]	Effective Dates	Basis for Import Allowance Requirement	Adjustments to Import Allowances	Allowance Allocation for Transitional Assistance[b]
S. 1766	Primary Products	2020–2050	Increase over Baseline in a Covered Good's Embodied Carbon Emissions	Based on Transitional Assistance and Economic Development of Exporting Country	10% of Allowances Allocated to Energy-Intensive Manufacturing in 2012 and Phased out by 2043
S. 3036	Primary Products	2020–2050	Increase over Baseline in a Covered Good's Embodied Carbon Emissions	Based on Transitional Assistance and Economic Development of Exporting Country	10% of Allowances Allocated to Energy-Intensive Manufacturing in 2012 and Phased out by 2030
H.R. 6186	"Trade-Exposed" Primary than fuel)	2020–2050	A Covered Good's Embodied Carbon Emissions Goods (Other 2020–2050	Based on Economic Development of Exporting Country	6% of Allowances Allocated to Energy-Intensive Manufacturing from 2012 to 2019
H.R. 6316	Primary Products and Manufactured Items for Consumption	2015–2050	A Covered Good's Embodied Carbon Emissions	Based on Transitional Assistance, Degree of Comparable Effort, and Economic Development of Exporting Country	10% of Allowances Allocated to Energy-Intensive Manufacturing in 2012 and Phased out by 2019

Source: Legislative Information System (LIS), available at www.congress (last accessed September 4, 2008).

a. A "covered good" is a primary product (or a manufactured item for consumption in the case of H.R. 6316) that generates, in the course of its manufacture, a substantial quantity of direct or indirect greenhouse-gas emissions and whose cost of production in the United States would be affected by the nation's climate policy; a "primary product" is defined as iron, steel, aluminum, or other manufactured product that is sold in bulk for the purpose of further manufacture (see H.R. 6316, sec. 101); a "manufactured item for consumption" is defined as a good that is not a primary product but that generates a substantial amount of direct or indirect emissions attributable to the primary product(s) in the manufactured item.

b. "Transitional assistance" refers to allocations of domestic allowances free of charge (it can be linked to production decisions about such items as output or labor).

Another challenge is that the national origin of goods exported to the United States could be difficult to determine, particularly if willful schemes using importing and reexporting were used to obscure the origin of components produced in nations without a sanctioned climate policy.

The larger the scope of covered goods (and the greater the difference in border adjustments by countries with comparable climate policies), the greater the likelihood of such schemes.

Transitional Assistance

Recent proposals would provide transitional assistance to energy-intensive manufacturers by giving allowances to firms in certain industries: 10 percent of the allowances beginning in 2012 under S. 1766, S. 3036, and H.R. 6316, and 6 percent of the allowances beginning in 2012 under H.R. 6186. Recent proposals would phase out transitional assistance more quickly than earlier proposals—H.R. 6186 and H.R. 6316 would cease allocations to energy-intensive manufacturing in 2019.

Giving allowances to energy-intensive manufacturers for free would not, in general, change their responses to a climate policy unless the grants were explicitly tied to specific production decisions. S. 1766, S. 3036, and H.R. 6316 would tie the amount of allowances allocated to an individual firm to the number of employees in that company relative to the sector average—providing an incentive to increase production or to substitute labor for capital or energy. H.R. 6186 would tie the allocation of transitional assistance to the output of an individual firm relative to the average output in the sector, thus providing an incentive for increased production. In those cases, transitional assistance in the short run could help energy-intensive manufacturers maintain their output and thereby limit the loss of U.S. jobs. Over the longer term, regardless of how the allowances were allocated, many of the affected sectors would probably be able to offset higher energy costs by substituting fuels containing less carbon and adopting technological advances.[9]

Compatibility with WTO Agreements

Adopting border adjustments would raise questions about whether using them to reduce carbon leakage would be compatible with U.S. obligations under the various World Trade Organization (WTO) agreements. Border adjustments might be challenged by other member countries of WTO. Even if such border adjustments were eventually found to be legal, many years could pass before rulings by WTO dispute settlement panels and the WTO Appellate Body (in cases challenging not only the U.S. program but other countries' programs as well) made clear precisely what was and what was not allowed. Only a cursory survey of the issues is possible here.

Each member country has agreed to upper limits, known as tariff bindings, beyond which it is not allowed to raise its tariffs. Almost all U.S. tariffs are at or near their limits, so border adjustments or other charges on imports under a cap-and-trade policy would be illegal under the WTO agreements unless covered by one of the exceptions contained in the agreements. Recent analyses point to two exceptions that the United States might use as defenses in any challenge to a border adjustment system.[10]

First, Articles II and III from the General Agreement on Tariffs and Trade (GATT— one of the WTO agreements) allow the imposition of taxes on imports that are equivalent to internal taxes imposed on like domestic goods. Several issues would arise in such a defense, not the least of which is whether two otherwise identical goods were "like" within the meaning of the WTO agreement and jurisprudence if the amount of CO_2 emitted in their production was different. In the past, determinations of "like" goods have generally not considered the processes used to produce them. Moreover, the border adjustments on imports would have to adhere to the most-favored-nation requirements of the agreement, which require that imports from all countries be treated the same. That requirement might mean that the border adjustments could not be reduced for imports from countries taking measures to reduce CO_2 emissions.

The second exception that the United States might use as a defense in a challenge is described in GATT Article XX. Paragraph (g) of that article allows for border measures "relating to the conservation of exhaustible natural resources if such measures are made effective in conjunction with restrictions on domestic production or consumption." A number of issues would arise in a defense based on that article as well. For example, past cases suggest that a dispute settlement panel would insist that there be a "sufficient nexus" between CO_2 emissions in the country challenging the tax and the consequences for the climate in the United States. If carbon leakage would be relatively small without a border adjustment (as some studies indicate would be the case), a dispute panel might conclude that the real reason for the border adjustment was not the prevention of carbon leakage but the protection of U.S. industry, which is not a valid justification under Article XX. The introduction to the article also requires that the "measures are not applied in a manner which would constitute a means of arbitrary or unjustifiable discrimination between countries where the same conditions prevail, or a disguised restriction on international trade." That phrase might be interpreted to require that border adjustments be reduced for imports from countries making efforts of their own to reduce emissions or for some developing countries.

Last, detailed sector or firm-level data on embodied carbon emissions could be needed to develop border adjustments because the WTO requires that producers from different countries be treated the same. Border adjustments based on sector averages have been found to be noncompliant in earlier WTO disputes.[11] If some or all of the allowances in the domestic cap-and-trade program were given away rather than sold, still other issues could arise. For example, they might be considered an actionable subsidy under the Subsidies and Countervailing Measures Agreement if a similar proportion of the permits required for imports were not given away.

End Notes

[1] Carbon sequestration is the capture and long-term storage of CO_2 emissions underground (geological sequestration) or in vegetation or soil (biological sequestration). For more information, see Congressional Budget Office, *The Potential for Carbon Sequestration in the United States* (September 2007).

[2] Those numbers are based on an analysis that CBO conducted using 1998 data; see Congressional Budget Office, *Who Gains and Who Pays Under Carbon-Allowance Trading? The Distributional Effects of Alternative Policy Designs* (June 2000). CBO is in the process of updating those figures, using recent data on households' expenditures and income.

[3] One researcher has suggested that an environmental tax credit based on earnings could offer another means of reducing the regressive effects of the price increases that would result from a tax or cap on CO_2 emissions. See Gilbert E. Metcalf, *A Proposal for a U.S. Carbon Tax Swap* (Washington, D.C.: Brookings Institution, October 2007).

[4] For those results, see Congressional Budget Office, *Trade-Offs in Allocating Allowances for CO_2 Emissions* (April 25, 2007).

[5] Specifically, CBO estimated that the value in 1998 of the allowances stemming from the 15 percent reduction in U.S. emissions would total $155 billion (in 2006 dollars). By comparison, profits for U.S. producers of oil, natural gas, and coal totaled $13.5 billion in 1998 (in 2006 dollars). Those companies' total profits have grown substantially—for example, in 2006, they totaled $174 billion.

[6] Although costs and benefits are difficult to measure, the long-term cumulative nature of climate change implies that the *benefit* of emitting one less ton of CO_2 in a given year—referred to as the marginal benefit—is roughly constant. In other words, the benefit in terms of averted climate damage from each additional ton of emissions reduced is roughly the same as the benefit from the previous ton of emissions reduced, and shifting the reductions from one year to another does not materially affect the ultimate impact on the climate. In contrast, the *cost* of emitting one less ton of CO_2 in a given year—the marginal cost—tends to increase with successive emission reductions. The reason is that the least expensive reductions are made first and progressively more-expensive cuts would then have to be made to meet increasingly ambitious targets for emission reductions.

[7] See William D. Nordhaus, "To Tax or Not to Tax: Alternative Approaches to Slowing Global Warming," *Review of Environmental Economics and Policy,* vol. 1, no. 1 (Winter 2007), pp. 37–39.

[8] Environmental Protection Agency, *Inventory of U.S. Greenhouse Gas Emissions and Sink: 1990–2006* (2008).

[9] See Morgenstern and others, *Competitiveness Impacts of Carbon Dioxide Pricing Policies on Manufacturing,* Resources for the Future, Climate Policy Forum, Issue Brief 7 (November 2007).

[10] Joost Pauwelyn, *U.S. Federal Climate Policy and Competitiveness Concerns: The Limits and Options of International Trade Law*, Working Paper NI WP 07-02 (Duke University, Nicholas Institute for Environmental Policy Solutions, April 2007); Jason E. Bordoff, "International Trade Law and the Economics of Climate Policy: Evaluating the Legality and Effectiveness of Proposals to Address Competitiveness and Leakage Concerns" (paper presented at the Brookings Institution's conference on Climate Change, Trade, and Competitiveness, Washington, D.C., June 9, 2008); Gary Clyde Hufbauer, Jisun Kim, and Steve Charnovitz, *Reconciling GHG Limits with the Global Trading System* (draft, Peterson Institute for International Economics, Washington, D.C., August 2008); Robert Howse and Antonia Eliason, "Domestic and International Strategies to Address Climate Change: An Overview of the WTO Legal Issues," in Bigdeli, Cottier, and Nartova, eds., *International Trade Regulation and the Mitigation of Climate Change* (Cambridge University Press, forthcoming); and Andrew Green, "Climate Change, Regulatory Policy and the WTO: How Constraining Are Trade Rules?" *Journal of International Economic Law,* vol. 8, no. 1 (2005).

[11] See, for example, Appellate Body, United States—Standards for Reformulated and Conventional Gasoline, WT/DS2/AB/R (April 29, 1996).

In: Policy Option Issues for CO2 Emissions ISBN: 978-1-60741-381-3
Editors: Nikolaus Vogler

Chapter 3

POLICY OPTIONS FOR REDUCING CO_2 EMISSIONS

Congressional Budget Office

SUMMARY

Global climate change is one of the nation's most significant long-term policy challenges. Human activities are producing increasingly large quantities of greenhouse gases, particularly carbon dioxide (CO_2), which accumulate in the atmosphere and create costly changes in regional climates throughout the world. The magnitude of such damage remains highly uncertain, but there is growing recognition that some degree of risk exists for the damage to be large and perhaps even catastrophic. Reducing greenhouse-gas emissions would be beneficial in limiting the degree of damage associated with climate change. However, decreasing those emissions would also impose costs on the economy—in the case of CO_2, because much economic activity is based on fossil fuels, which release carbon in the form of carbon dioxide when they are burned. Most analyses suggest that a carefully designed program to begin lowering CO_2 emissions would produce greater benefits than costs.

The most efficient approaches to reducing emissions involve giving businesses and individuals an incentive to curb activities that produce CO_2 emissions, rather than adopting a "command and control" approach in which the government would mandate how much individual entities could emit or what technologies they should use. Incentive-based policies include a tax on emissions, a cap on the total annual level of emissions combined with a system of tradable emission allowances, and a modified cap-and-trade program that includes features to constrain the cost of emission reductions that would be undertaken in an effort to meet the cap. In this study, the Congressional Budget Office (CBO) compares these incentive-based approaches, focusing on three key criteria:

- Efficiency in maintaining a balance between the uncertain benefits and costs of reducing CO_2 emissions,
- Ease or difficulty of implementation, and
- Possible interactions with other countries' policies for curbing CO_2—that is, the potential to ensure that U.S. and foreign policies produce similar incentives to cut emissions inside and outside the United States.

Other criteria could be of interest to policymakers in determining how best to address concerns about climate change. For example, the efficiency criterion addresses how well policies might function to minimize the cost of reducing emissions over a period of several decades; however, policymakers may choose to place more emphasis on providing certainty about the amount of emissions at specific points in time. Similarly, policymakers may also wish to focus on how different policy designs affect different segments of society.

Policy Options for Reducing Emissions

Incentive-based approaches can reduce emissions at a lower cost than more restrictive command-and-control approaches because they provide more flexibility about where and how emission reductions are achieved.

Under a tax, policymakers would levy a fee for each ton of CO_2 emitted or for each ton of carbon contained in fossil fuels. The tax would motivate entities to cut back on their emissions if the cost of doing so was less than the cost of paying the tax. As a result, the tax would place an upper limit on the cost of reducing emissions, but the total amount of CO_2 that would be emitted in any given year would be uncertain.

In contrast, under a cap-and-trade program, policy-makers would set a limit on total emissions during some period and would require regulated entities to hold rights, or allowances, to the emissions permitted under that cap. (Each allowance would entitle companies to emit one ton of CO_2 or to have one ton of carbon in the fuel that they sold.) After the allowances for a given period were distributed, entities would be free to buy and sell the allowances among themselves. Unlike a tax, a cap-and-trade program would place an upper limit on the amount of emissions, but the cost of reducing emissions would vary on the basis of fluctuations in energy markets, the weather (for example, an exceptionally cold winter would increase the demand for energy and make meeting a cap more expensive), and the technologies available for reducing emissions.

Given the gradual nature of climate change, the uncertainty that exists about the cost of reducing emissions, and the potential variability of the cost of meeting a particular cap on emissions at different points in time, a tax could offer significant advantages. If policymakers chose to specify a long-term target for cutting emissions, a tax could be set at a rate that could meet that target at a lower cost than a comparable cap. In addition, if policymakers set the tax rate at a level that reflected the expected benefits of reducing a ton of emissions (which would rise over time), a tax would keep the costs of emission reductions in balance with the anticipated benefits, whereas a cap would not.

There is significant interest, however, in a cap-and-trade approach (which has been used in the United States to reduce emissions that cause acid rain and is currently being used in the European Union to limit CO_2 emissions).[1] This study therefore explores ways in which policymakers could preserve the structure of a cap-and-trade program but achieve some of the efficiency advantages of a tax. Specifically, policymakers could take one or more of these steps:

- Set a ceiling—typically referred to as a safety valve—or a floor on the price of emission allowances. The government could maintain a ceiling by selling companies as many allowances as they would like to buy at the safety-valve price. The government could maintain a price floor by selling a significant fraction of allowances in an auction and specifying a reserve price.
- Permit firms to transfer emission-reduction requirements across time—by "banking" allowances in one year for use in future years or by "borrowing" future allowances for use in an earlier year. Firms would have an incentive to bank allowances when the cost of cutting emissions was low (relative to anticipated future costs) and to borrow allowances when costs were high.

- Modify the stringency of the cap from year to year on the basis of the price of allowances. Policymakers could loosen the cap if the price of allowances rose too high, or they could tighten the cap if the price fell too low. Some analysts have suggested the use of a "circuit breaker" that would halt the gradual tightening of the cap if the price of allowances exceeded a specified trigger price. The cap would resume its decline if the price of allowances eventually fell below the trigger price. Loosening or tightening the cap could be achieved indirectly by altering conditions under which firms could bank or borrow allowances.

RESULTS OF CBO'S ANALYSIS

The policy options described above differ in their potential to reduce emissions efficiently, to be implemented with relatively low administrative costs, and to create incentives for emission reductions that are consistent with incentives in other countries. CBO draws the following conclusions:

- A tax on emissions would be the most efficient incentive-based option for reducing emissions and could be relatively easy to implement. If it was coordinated among major emitting countries, it would help minimize the cost of achieving a global target for emissions by providing consistent incentives for reducing emissions around the world. If other major nations used cap-and-trade programs rather than taxes on emissions, a U.S. tax could still provide roughly comparable incentives for emission reductions if the tax rate each year was set to equal the expected price of allowances under those programs. (See Summary Table 1 for a qualitative comparison of selected policies.)
- An inflexible annual cap (one whose level was not affected by the price of emission allowances and under which firms would not be allowed to bank or borrow allowances) would be the least efficient option among those considered here, although it could be relatively easy to implement, depending on key design features. Linking the cap-and-trade programs of various countries could create significant concerns, however: Nations would give up sovereignty over the price of the allowances traded in their programs and the extent to which emissions were reduced in ways that met their programs' criteria.

- A cap-and-trade program that included a price ceiling (safety valve) and either a price floor or banking provisions could be significantly more efficient than an inflexible cap, although somewhat less efficient than a tax. It might also be relatively easy to implement, depending on specific design decisions. If major emitting countries agreed to establish such programs—and to set their safety valves at roughly the same level— they could create similar incentives to reduce emissions without formally linking their cap-and-trade programs. Alternatively, if other developed countries taxed CO_2 emissions, a safety valve in a U.S. cap-and-trade program could be set at a level consistent with that tax.
- Moderating the price of allowances by altering the stringency of a cap—or the extent to which firms could use banked and borrowed allowances—would be considerably more difficult to implement than setting a price floor or ceiling directly. Price volatility in the allowance market could make it difficult for policymakers to know when to alter the supply of allowances and would mean that no particular price outcome could be guaranteed. One particular form of price-sensitive cap—a cap-and-trade program with a circuit breaker—could be more efficient than an inflexible cap. However, such a program would be less efficient than the other policy options that CBO examined.

Comparison of Policies' Efficiency

The most efficient policy tool for decreasing CO_2 emissions is the one that can best balance the costs and benefits of the reductions, even when both are uncertain. The features that make a policy tool most efficient would also enable it to minimize the cost of achieving a given target, even if that target was not explicitly chosen to balance costs and benefits.

A Tax Versus an Inflexible Cap

Analysts generally conclude that a tax would be a more efficient method of reducing CO_2 emissions than an inflexible cap. The efficiency advantage of a tax stems from the contrast between the long-term cumulative nature of climate change and the short-term sensitivity of the cost of emission reductions. Climate change results from the buildup of CO_2 in the atmosphere over several decades; emissions in any given year are only a small portion of that total. As a result, limiting climate change would require making substantial reductions in those emissions over many years, but ensuring that any *particular limit* was met in any *particular year* would result in little, if any, additional benefit (avoided damage). In contrast, the cost of cutting emissions by a particular amount in a given year could vary significantly depending on a host of factors, including the weather, disruptions in energy markets, the level of economic activity, and the availability of new low-carbon technologies (such as improvements in wind-power technology).

Relative to a cap-and-trade program with prespecified emission limits each year, a steadily rising tax could better accommodate cost fluctuations while simultaneously achieving a long-term target for emissions. Such a tax would provide firms with an incentive to undertake more emission reductions when the cost of doing so was relatively low and allow them to reduce emissions less when the cost of doing so was particularly high. In contrast, an inflexible cap-and-trade program would require that annual caps were met regardless of the cost, thereby failing to take advantage of low-cost opportunities to cut more emissions than were required by the cap and failing to provide firms with leeway in years when costs were higher.

The efficiency advantage of a tax over an inflexible cap depends on how likely it is that actual costs will differ from what policymakers anticipated when they set the level of the cap. Given the uncertainties involved, such differences are likely to be large—and, therefore, analysts generally conclude that the efficiency advantage of a tax is likely to be quite large. Specifically, available research suggests that in the near term, the net benefits (benefits minus costs) of a tax could be roughly five times greater than the net benefits of an inflexible cap.[2] Put another way, a given long-term emission-reduction target could be met by a tax at a fraction of the cost of an inflexible cap-and-trade program.

Flexible Cap Approaches

A cap-and-trade program could incorporate various design features that would keep allowance prices from rising or falling farther than policymakers wanted. Combined, some of those features could allow a cap-and-trade program to achieve many of the efficiency advantages of a tax on emissions.

Summary Table 1. Comparison of Selected Policies for Cutting CO_2 Emissions

	Efficiency		Implementation Considerations	International Consistency Considerations
Policy	**Ranking**	**Considerations**		
Carbon Dioxide Tax	1	A tax would avoid significant year-to-year fluctuations in costs. Setting the tax equal to the estimate of the marginal benefit of emission reductions would motivate reductions that cost less than their anticipated benefits but would not require reductions that cost more than those benefits. Research indicates that the net Benefits of a tax could be roughly five times as high as the net benefits of an inflexible cap. Alternatively, a tax could achieve a long-term target at a fraction of the cost of an inflexible cap.	An upstream tax would not require monitoring emissions and could be relatively easy to implement. It could build on the administrative infrastructure for existing taxes, such as excise taxes on coal and petroleum.	A U.S. tax could be set at a rate consistent with carbon dioxide taxes in other countries. Consistency would require comparable verification and enforcement. If countries imposed taxes at different points in the carbon supply chain, special provisions could be needed to avoid double-taxing or exempting certain goods. Setting a U.S. tax that would be consistent with allowance prices under other countries' cap-and-trade systems would be somewhat more difficult because it would require predicting allowance prices in different countries.
Cap With Safety Valve and Either Banking or a Price Floor	2	A cap-and-trade program that included a safety valve and either banking or a price floor could have many of the efficiency advantages of a tax. The safety valve would prevent price spikes and could keep the costs of emission reductions from exceeding their expected benefits. Banking would help prevent the price of allowances from falling too low,	An upstream cap would not require monitoring emissions. It would require a new administrative infrastructure to track allowance holdings and transfers. Implementing a safety valve would be straightforward: The government would offer an unlimited number of allowances at the safety-valve price.	Either a safety valve or banking would become available to all sources of CO2 emissions in a linked international cap-and-trade program. Some countries could object to linking with a U.S. program that included those features, because linked countries could not ensure that their emissions would be below a required level in a given year. Linking would also create concerns about inconsistent monitoring and enforcement among countries and international capital flows (as described

Summary Table 1. (Continued)

	Efficiency		Implementation Considerations	International Consistency Considerations
Policy	Ranking	Considerations		
		provided that prices were expected to be higher in the future. A price floor, however, would be more effective at keeping the cost of emission reductions from falling below a target level.	Banking has been successfully implemented in the U.S. Acid Rain Program. A price floor would be straightforward to implement only if the government chose to sell a significant fraction of emission allowances in an auction.	below in the inflexible cap policy).Countries with different cap-and-trade programs could capture many of the efficiency gains that would be achieved by linking—while avoiding some of the complications—if they each included banking (or set a similar price floor) and agreed on a safety-valve price.
Cap With Banking and Either a Circuit Breaker or Managed Borrowing	3	Allowing firms to bank allowances would help prevent the price of allowances from falling too low, provided that prices were expected to be higher in the future. Including a circuit breaker—or increasing the ability of firms to borrow allowances—would help keep the price of allowances from climbing higher than desired, but would be significantly less effective at doing so than a price ceiling.	An upstream cap would not require monitoring emissions. It would require a new administrative infrastructure to track allowance holdings and transfers. Banking has been successfully implemented in the U.S. Acid Rain Program. Determining when to trigger a circuit breaker, or modify borrowing restrictions, would require judgment about current and future allowance prices. Such interventions could aggravate price fluctuations if those judgments were incorrect.	Including banking and either a circuit breaker or borrowing in the U.S. program could reduce the likelihood of linking because it would cause uncertainty about the stringency of the U.S. cap relative to other countries' caps and about the total supply of allowances in the global trading market.

	Efficiency		Implementation Considerations	International Consistency Considerations
Policy	Ranking	Considerations		
Inflexi-ble Cap	4	Allowance prices could be volatile. An inflexible cap could require too many emission reductions (relative to their benefits) if the cost of achieving them was higher than anticipated and could require too few reductions if the cost of meeting the cap was lower than policymakers had anticipated.	An upstream cap would not require monitoring emissions. It would require a new administrative infrastructure to track allowance holdings and transfers.	Linking an inflexible U.S. cap with other countries' cap-and-trade systems would create a consistent global incentive for reducing emissions. However, inconsistent monitoring and enforcement in any one country could undermine the entire linked trading system. Further, linking would alter allowance prices in participating countries, create capital flows between countries, and possibly encourage countries to set their caps so as to influence those flows.

Source: Congressional Budget Office.

Note: An "upstream" tax or cap would be imposed on suppliers of fossil fuel on the basis of the carbon dioxide (CO_2) emitted when the fuel was burned. A "safety valve" would set a ceiling on the price of allowances. "Banking" would allow firms to exceed their required emission reductions in one year and use their extra allowances in a later year. Under a "circuit breaker," the government would stop a declining cap from becoming more stringent if the price of allownces exceeded a specified level.

Keeping Costs From Climbing Too High

Including a safety valve could make a cap-and-trade program more efficient than an inflexible cap. Such a policy would set a ceiling on the price of allowances, preventing the cost of reducing emissions from exceeding either the best available estimate of the benefit (avoided damage) that would result from those reductions or the cost that policymakers consider acceptable.

Alternatively, policymakers could attempt to cap the price of allowances by adjusting the stringency of the cap. For example, policymakers could specify a circuit breaker, which would prevent a declining cap from becoming more stringent (fixing the cap at one level) if the price of allowances reached a certain level. Unlike a safety valve, a circuit breaker would not necessarily stop the price of allowances from continuing to rise, but it would result in smaller price increases than would otherwise occur. (The price would probably still increase because meeting a fixed cap would become more and more costly over time as the economy grew.)

Finally, allowing companies to borrow allowances—and thus defer emission reductions to the future—could help keep the price of allowances from rising too high. Policy-makers could alter the constraints placed on firms' use of borrowed allowances on the basis of the price of allowances. Like a circuit breaker, such an approach could help constrain the price of allowances under some circumstances, but it is unlikely to be as effective at doing so as a safety valve. Policymakers would need to forecast future allowance prices in order to know when to loosen or tighten constraints on borrowing. To the extent that those forecasts were inaccurate, borrowing could exacer-bate price fluctuations. Further, firms would find it profitable to borrow future allowances only if they expected the price of allowances to be lower in the future. That is, borrowing could help deal with temporary spikes in allowance prices but not circumstances in which allowance prices were expected to remain high in the long term.

Keeping Costs From Falling Too Low

Policymakers could prevent the price of allowances from falling too low by setting a price floor. If the government chose to sell a significant portion of the allowances by auction, it could specify a reserve price and withhold allowances from the auction as needed to maintain that price. Attempting to prevent the price of allowances from dropping too low by adjusting the supply of allowances would entail the same complications associated with a circuit breaker.

Alternatively, policymakers could help keep the price of allowances from falling below some desired level by allowing companies to exceed their required emission reductions in low-cost years in order to bank allowances for use in future high-cost years. The additional emission reductions motivated by banking in low-cost years would put upward pressure on the price of allowances in those years. Similarly to borrowing, banking would be most effective in addressing short-term lows in allowance prices rather than circumstances in which allowance prices were expected to remain low in the long term.

Comparison of Policies' Implementation

Policies that are efficient in theory will be efficient in practice only if they can be implemented effectively with-out excessive administrative costs. Either a tax or an inflexible cap could meet that criterion.

Administering an "upstream" tax or cap-and-trade program for CO_2 emissions would involve taxing or regulating the suppliers of fossil fuels—such as coal producers, petroleum refiners, and natural gas processors. Compared with a "downstream" design, which would tax or regulate users of fossil fuels, an upstream approach would have two administrative advantages. It would involve regulating a limited number of entities, and it would not require firms to monitor actual emissions. Rather, each firm's tax payment or allowance requirement could be based on the carbon content of its fuel and the amount it sold.[3]

An upstream tax may be somewhat easier to implement than an upstream cap-and-trade program because many of the entities that would be covered by either policy are already subject to excise taxes.[4] A CO_2 tax could build on that existing structure. Implementing a cap-and-trade program, by contrast, would probably require a new administrative infrastructure. However, the Environmental Protection Agency's experience with the Acid Rain Program (a cap-and-trade program designed to reduce emissions of sulfur dioxide by electricity generators) suggests that the cost of administering such a program could be modest.

Some design features that might improve the efficiency of a cap-and-trade program—such as a price ceiling, banking, and borrowing—could be implemented without unduly increasing administrative costs. A price floor could be relatively easy to implement, but only if the government chose to auction off a significant fraction of the allowances. Other design features could prove more challenging to implement. For example, determining the basis for triggering a circuit breaker (or, more generally, for loosening or tightening the stringency of a cap) would require the government to make judgments about current and future allowance prices.

Comparison of Policies' International Consistency

Carbon dioxide is a global pollutant. A ton of emissions from any point on the globe at any given time would have the same effect on the atmospheric concentration of CO_2 and thus would cause the same amount of damage.

Consequently, the most costeffective way to reach a specific atmospheric concentration would be to undertake the lowest-cost emission reductions regardless of where they were located. Achieving that goal would require creating a uniform incentive to reduce emissions in countries that are major emitters of CO_2.

One option is to have each of the major emitting countries agree to adopt a similar tax on CO_2 emissions. However, a system of harmonized taxes would produce a consistent global incentive for cutting emissions only if participating countries also adopted similar monitoring, verification, and enforcement provisions.

Alternatively, major emitting nations could agree to link their cap-and-trade programs. In that case, competitive forces would equalize the price of allowances between countries and create consistent incentives to reduce emissions. Uniformity of monitoring and enforcement would be even more important in such an international program. With harmonized taxes, lax monitoring or enforcement by any one country could reduce the incentives for emission reductions in that country. But with linked cap-and-trade programs, laxity in one area could undermine the integrity of allowances throughout the entire system. In addition, linking existing cap-and-trade programs could result in significant flows of capital between countries (from the sale of allowances) and could encourage a nation to set the level of its cap so as to influence those flows.

If the United States included a safety valve or banking or borrowing provisions in its cap-and-trade program, those design features would become available to all sources of CO_2 emissions within a linked cap-and-trade system, regardless of their location. The increased flexibility provided by those design features could undermine the ability of all participating countries to meet a fixed emissions limit in a given year or compliance period; thus, they could be seen as an obstacle to linking with a U.S. cap-and-trade program. For example, if the United States had a cap-and-trade program with a safety valve and linked that program to the European Union's Emission Trading Scheme, which has a fixed cap and no safety valve, countries in the European system would no longer be able to ensure that they could meet the fixed caps they agreed to under the Kyoto Protocol.

Alternatively, any set of policies that resulted in a similar allowance price in different countries would produce efficiency gains similar to those of linking, without requiring nations to give up sovereignty over the price of their allowances or the integrity of their programs. For example, countries with nonlinked cap-and-trade programs could agree to include a safety valve set at a similar level, or the United States could set its safety valve at the same level as a CO_2 tax in another country.

One challenge in crafting an efficient global approach to cutting CO_2 emissions is the inclusion of developing countries that are becoming (or are expected to become) major emitters. China, for example, contributed roughly 8 percent of the world's CO_2 emissions from fossil fuels in 1980, but its share reached 19 percent in 2005. (During the same period, the U.S. share of global emissions fell from 26 percent to 21 percent.[5]) Some researchers suggest that a system of linked cap-and-trade programs could equalize the marginal cost of emission reductions among participating countries while allowing for different levels of reduction among the countries on the basis of fairness or other criteria.[6] Alternatively, some analysts suggest that the revenue generated by taxing CO_2 emissions or selling emission allowances in developed coun tries could be used to fund emission reductions in developing nations.[7]

Other opportunities also exist for including developing countries. For example, in the European Union's trading program for CO_2 emissions, companies are allowed to comply with some of their allowance requirements by funding emission reductions in developing countries, such as financing a low-emission power plant in China.

1. Efficiency Implications of Different Policy Designs

Incentive-based policies can reduce emissions of carbon dioxide (CO_2) and other greenhouse gases, thereby reducing the risks associated with global climate change, at a lower cost than less flexible alternatives. Policymakers have many options, however, for giving businesses and households an economic incentive to reduce emissions. One option is to regulate the price of emissions—for example, by imposing a tax on them. A tax would limit the cost of cutting emissions but would leave the amount of CO_2 emitted in a given year uncertain. As an alternative, the government could adopt a market-based system to regulate the quantity of emissions—for instance, by combining a cap on total annual emissions with a system of tradable emission permits, or allowances. If monitoring and enforcement were effective, a cap-and-trade program would limit the amount of CO_2 emitted in a given year but would leave the cost of reducing emissions uncertain. The design of a cap could be modified in various ways to make it more flexible and to adopt some of the characteristics of a tax while maintaining the structure of a cap-and-trade program.

Any of those incentive-based approaches could achieve a given cut in emissions at a lower cost than command-and-control approaches, in which the government mandated how much individual factories could emit or what technologies they should use. However, incentive-based approaches would differ in their economic efficiency (the subject of this chapter) and in the ease with which they could be implemented in the United States and coordinated with other countries' emission-reduction policies (discussed in Chapters 2 and 3). The most economically efficient policy is the one that can best keep the marginal cost of reducing emissions—that is, the cost of cutting emissions by another ton—in balance with the marginal benefit (in terms of avoided damage from climate change). A related concept is cost-effectiveness. A cost-effective policy would minimize the cost of meeting a given target for emissions, regardless of whether or not that target was chosen to balance benefits and costs. The efficiency criterion addresses how well policies might function to minimize the cost of reducing emissions over a period of several decades; however, policymakers may choose to place more emphasis on providing certainty about the amount of emissions at specific points in time.

Neither the costs nor the benefits of reducing CO_2 emissions can be known when a reduction policy is put in place. Thus, policymakers must rely on estimates of both of them. The costs of reducing emissions would occur when the reductions were made and could vary substantially depending on such factors as the amount of economic activity, market conditions, weather, and available technologies. The benefits of reducing emissions, in contrast, would be realized decades or even centuries after the reductions were made. The reason is that each ton of CO_2 generates a rise in the average global temperature that peaks about 40 years after the CO_2 is emitted and then dissipates slowly, with a half-life of about 60 years.[8]

Estimating the benefits of cutting emissions is complicated by that long-term effect. In addition, analysts who try to estimate the benefits of cutting emissions face many other challenges, including addressing numerous scientific and economic uncertainties; measuring costs, such as mass species extinction, that are difficult to quantify in economic terms; and deciding how much weight to give to changes in the welfare of future generations.[9]

Some experts think that the effects of climate change could be modest, especially if society is ingenious in adapting to the change. However, other experts are concerned that rising concentrations of greenhouse gases could produce far more severe consequences for the global and U.S. economies than have generally been projected. Curbing greenhouse-gas emissions would help limit not only the expected costs of future global climate change but also the chances of irreversible or potentially catastrophic damage.

In general, the possibility of significant damage provides an economic motivation for taking additional action to moderate the growth of emissions in the near future—and, potentially, to cut emissions to very low levels in the longer run. Individuals take actions (such as reducing risky behavior or buying insurance) to lessen their harm from extreme events; similarly, societies or governments should and do take actions to avoid catastrophic collective harm. The difficulty for policymakers is determining the appropriate cost to be paid today to lessen what may be a small risk of a potentially catastrophic event in the future.[10]

Although estimating the benefits of emission reductions is difficult, policymakers cannot avoid making a judgment about them: Policy choices about climate change will necessarily imply a value for those benefits. That value would be explicit under a tax, because the tax rate provides an indication of what the government thinks an incremental reduction in emissions is worth. By contrast, that value would be implicit under a cap. A higher (less stringent) cap would imply a lower estimate of the marginal benefit of cutting emissions—as reflected in lower prices for emission allowances—than a lower (more stringent) cap would.

When comparing emission-reduction policies, the Congressional Budget Office (CBO) generally assumes that lawmakers would design them in the most efficient way—that is, to achieve the highest possible net benefits, given the limitations of each particular policy tool. Thus, for example, this analysis compares the most efficient tax on CO_2 with the most efficient cap. In other words, the tax or cap is assumed to be set at a level that encourages the affected parties to reduce emissions as long as the expected cost of doing so is less than or equal to the expected benefit. Those costs and benefits will inevitably be different than anticipated. Policy designs will yield different net benefits depending on their ability to balance the costs and benefits of emission reductions when those turn out to be higher or lower than policymakers had anticipated. Designs that are relatively more efficient would also be relatively cost-effective: The characteristics of a policy design that enable it to equate the cost of additional emission reductions with their anticipated benefits also enable it to minimize the cost of achieving any given emission-reduction target.

To be most efficient, a tax would need to rise and a cap would need to decline gradually over time. The future benefits of avoiding climate-change damage by reducing CO_2 emissions by a ton would have an increasingly greater present value (that is, the value today after taking into account the time value of money) as the potential for large damage drew closer in time. An increasingly stringent tax or cap would reflect that increase in present value over time. Further, a gradually rising tax or tightening cap would allow for a smoother transition to a less carbon-intensive economy. Businesses and households would have more time to replace their equipment and energy-use practices with more efficient alternatives.

A Carbon Dioxide Tax versus an Inflexible Carbon Dioxide Cap

According to many analysts, a tax would be a more economically efficient policy for reducing CO_2 emissions than an inflexible cap (with "inflexible" meaning a cap whose level was not affected by the price of emission allowances). That conclusion stems from the cumulative, long-term nature of climate change: The benefit of emitting one less ton of CO_2 in a given year is roughly constant, whereas the cost of emitting one less ton of CO_2 each year rises with each ton reduced. The reason for rising marginal costs is that companies that have to comply with an emission-reduction policy will make the cheapest cuts first and progressively more expensive cuts thereafter.

The contrast between constant marginal benefits and rising marginal costs means that the gap between uncertain costs and benefits is particularly sensitive to the amount of annual emission reductions. A cap that is too tight will disproportionately increase costs over benefits, and a cap that is not tight enough will disproportionately lower costs relative to benefits. A tax, by contrast, will tend to hold the costs of emission reductions in line with the constant (although uncertain) expected benefits, encouraging greater emission reductions when costs are low and allowing more emissions when costs are high.

An Illustrative Example of How a Tax Would Be More Efficient Than a Cap

To understand how a tax could offer efficiency advantages over a cap, assume that the future benefits of limiting emissions have a present value of \$15 per metric ton of CO_2 (or \$55 per metric ton of carbon), that those bene-

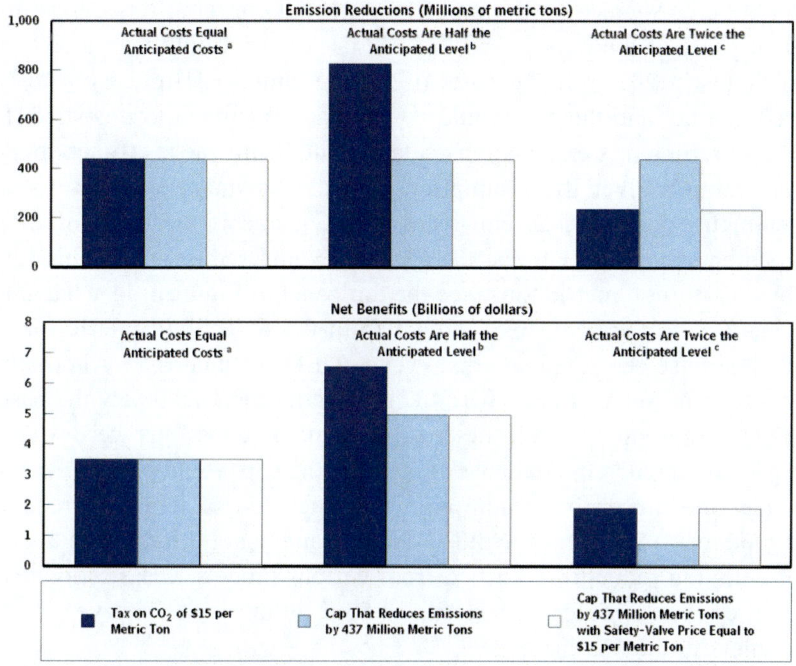

Source: Congressional Budget Office.

Notes: For illustrative purposes only, this example assumes that the benefit of reducing carbon dioxide (CO_2) emissions is $15 per metric ton. It examines the net benefits that would result in the first year of each policy, assuming that the policy covered only the United States and took effect in 2017 after having been announced 10 years earlier. The cost of firms' emission reductions (and the response to various taxes) is derived from Mark Lasky, *The Economic Costs of Reducing Emissions of Greenhouse Gases: A Survey of Economic Models*, Congressional Budget Office Technical Paper No. 2003-03 (May 2003).

A safety valve is a ceiling on the price of emission allowances.

a. Assumes that the actual marginal cost of reducing emissions by 437 million metric tons is $15 per metric ton, the cost that policymakers anticipated when they set the cap.

b. Assumes that the actual marginal cost of reducing emissions by 437 million tons is $7.50 per metric ton but that the tax induces more reductions (up to 824 million tons) at a marginal cost of $15 per metric ton.

c. Assumes that the actual marginal cost of reducing emissions by 437 million tons is $30 per metric ton but that the tax induces fewer reductions (234 million tons instead of 437 million), up to a marginal cost of $15 per metric ton.

Figure 1-1. Illustrative Comparison of Various Policies to Reduce CO_2 Emissions Under Different Cost Conditions

fits would be constant over the range of potential emission reductions during the initial years of the policy, and that the tax or cap would take effect in the United States in 2017.[11] If the costs of cutting emissions turned out to be as expected, the tax and the cap would be equivalent. But if those costs differed from the government's expectations, a tax would be the more efficient policy.

For example, given the assumptions above, if lawmakers imposed a tax of \$15 per metric ton on U.S. emissions of CO_2, and if the costs of limiting emissions equaled expectations, the \$15 tax would reduce U.S. emissions in 2017 by 437 million metric tons (see the top panel of Figure 1-1). That amount represents a cut of roughly 6.5 percent from the 6.7 billion metric tons that would otherwise be emitted that year, CBO estimates.[12] Alternatively, lawmakers could set a cap that was 437 million metric tons below the baseline level of U.S. emissions, and if the costs of reducing emissions were what they had expected, the incremental cost of meeting the cap would be \$15 per metric ton. Under the illustrative assumption that each ton of emission reductions would produce \$15 worth of avoided damage and using information about the cost of emission reductions derived from various models, CBO estimates that either policy would yield net benefits of \$3.5 billion in its first year (see the lower panel of Figure 1-1).[13]

If the costs of cutting emissions were different than expected, however—for example, if new technologies turned out to be less expensive than anticipated—the two policies would produce different outcomes.

If the costs of cutting emissions were half the anticipated level—for example, because of unforeseen technological breakthroughs—both policies would produce higher net benefits than expected.[14] The increase in net benefits, though, would be greater under a tax than under a cap: The tax would give firms an incentive to keep cutting emissions as long as doing so cost less than paying the tax. CBO estimates that in this scenario a tax would cause emissions to be cut by 824 million metric tons (roughly 12 percent below the baseline level), rather than by the 437 million metric tons required by the cap. Each of those additional cuts would boost net benefits because they would cost less than, or as much as, their \$15 per ton expected benefit.

Alternatively, if the cost of reducing emissions turned out to be twice as high as expected, the net benefits would be lower under each policy—but would fall much more under the cap than under the tax. In particular, under the inflexible cap, firms would be required to reduce emissions by 437 million metric tons, even though reaching that target would entail making reductions that cost up to \$30 per metric ton but provided benefits of only \$15 per metric ton. As a result of the higher costs, the total net benefits of the cap would fall to $0.7 billion—just one-fifth of the expected amount. A tax would also have lower net benefits if the costs of cutting emissions proved greater than expected. But net benefits would decline by less for a tax than for a cap. Because companies would have the flexibility to reduce emissions by less than 437 million metric tons, the net benefits of a tax would be more than twice those of a cap.

Like costs, benefits could also be higher or lower than anticipated; however, neither policy would adjust to that change. If actual marginal benefits turned out to be much higher than expected, either a tax or a cap would produce too few cuts in emissions, and both policies would fall short of the most efficient level of emission reductions by the same amount.[15]

Empirical Estimates of the Efficiency Advantage of a Tax

If the government wanted to maximize expected net benefits, it would need to set the level of a cap or a tax in a given year on the basis of its best estimate of both the costs and benefits of reducing emissions in that year. However, actual costs in any year are likely to differ from the best estimate, sometimes exceeding it and sometimes falling below it. Because a tax would motivate only emission reductions that cost less than the tax rate, it would automatically adjust the quantity of emission reductions to keep their costs in line with their anticipated benefits, whereas a cap would not.

When analysts take into account the degree to which costs are likely to vary around a single best estimate, they conclude that a tax could offer much higher net benefits than a cap. One study suggests that the net benefits of a worldwide tax on CO_2 emissions in 2010 would be more than eight times larger than those of an equivalent inflexible cap. If the policies are assumed to be set in place for 100 years, the efficiency advantage of a tax declines to a factor of five.[16] Another study concluded that a tax could offer up to 16 times greater expected net benefits than a cap under some assumptions.[17] A third study examined outcomes when cost shocks were assumed to be correlated across time—that is, an unusually high cost of meeting the cap in any given year increases the likelihood of a higher than average cost in the following year. Using their base-case parameter estimates for factors that might affect costs (such as baseline emissions and changes in technology) and assuming a 10-year policy, those researchers estimated that the net benefits of a tax would be roughly five times higher than those of a cap.[18] Taken together, those studies suggest that the net benefits of a tax could be roughly five times those of an inflexible cap (see Figure 1-2)—assuming that both policies were designed to balance expected costs and benefits.

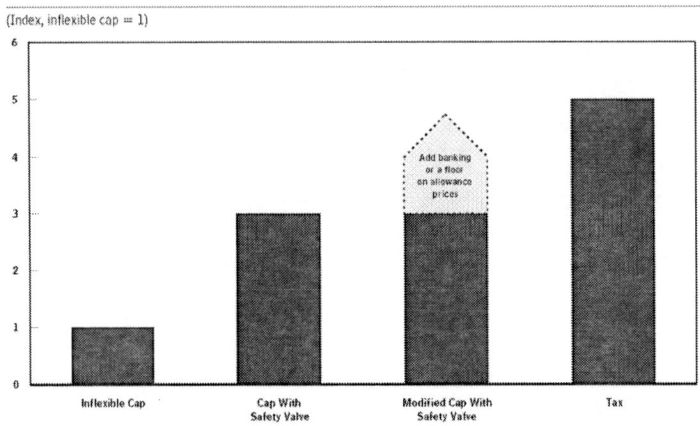

Source:Congressional Budget Office based on estimates of the relative magnitude of the net benefits of various policies found in William A. Pizer, "Combining Price and Quantity Controls to Mitigate Global Climate Change," *Journal of Public Economics*, vol. 85 (2002), pp. 409–434, and in Richard G. Newell and William A. Pizer, "Regulating Stock Externalities Under Uncertainty," Journal of Environmental Economics and Management, vol. 45 (2002), pp. 416–432.

Notes: The net benefits of each policy are shown in relationship to each other with the net benefits of an inflexible cap set equal to one. The inflexible cap and the tax are assumed to be set at the most efficient level—that is, at the point at which the expected marginal cost of complying with the policy would be equal to the anticipated marginal benefit of reducing emissions.

The net benefits of a cap with a safety valve (a ceiling on the price of emission allowances) are based on the assumption that the cap would be set at the level of the most efficient inflexible cap and the safety-valve price would be set at the level of the most efficient tax. Banking would enable firms to save unused allowances from one period to use in a future period.

The net benefits of a cap-and-trade program with a circuit breaker (not shown in the figure) would be greater than those of an inflexible cap and less than those of a cap with a safety valve; however, CBO lacked sufficient information to determine how much greater or less they would be.

A cap-and-trade program that included a safety valve and either a price floor or banking provisions could be significantly more efficient than an inflexible cap, although somewhat less efficient than a tax.

CO_2 = carbon dioxide.

Figure 1.2. Relative Economic Efficiency of Various Policies to Reduce CO_2 Emissions, When Cost Uncertainty Is Taken Into Account.

Viewed another way, any long-term emission-reduction target could be met by a tax at a fraction of the cost of an inflexible cap-and-trade program. That cost savings stems from the fact that a tax could better accommodate cost fluctuations while simultaneously achieving a long-term emission target. It would provide firms with an incentive to undertake more emission reductions when the cost of doing so was relatively low and allow them to reduce emissions less when the cost of doing so was particularly high.

The Impact of Price Volatility

The flexibility in reducing emissions that a tax affords is important because the cost of cutting emissions by a given amount could vary from year to year depending on such factors as the weather, the level of economic activity, and the availability of low-carbon technologies. A tax would provide a steady, predictable price for emissions. An inflexible cap, however, could result in volatile allowance prices, making a cap-and-trade program more disruptive to the economy than a tax would be.

Experience with cap-and-trade programs has shown that price volatility can be a major concern when a program's design does not include provisions to adjust for unexpectedly high costs and to prevent price spikes. For example, one researcher found that the price of sulfur dioxide allowances under the U.S. Acid Rain Program was significantly more volatile than stock prices between 1995 and 2006 (see Figure 1-3).[19]

Price volatility was most apparent in the summer of 2000 in Southern California's Regional Clean Air Incentives Market (RECLAIM), a program that capped emissions of nitrous oxide (NO_x) from the power sector. A heat wave caused demand for electricity to soar that summer, while the availability of imported power from other states declined. The increase in demand had to be met by running many of California's old gas-fired generating facilities, which had not yet installed NO_x emission controls.

As a result, the demand for NO_x RECLAIM Trading Credits for 2000 rose significantly, boosting their average annual price tenfold (from $4,284 per ton in 1999 to almost $45,000 per ton in 2000) and contributing to high wholesale electricity prices in California during that period.[20] In addition to the California experience, allowance prices in the European Union's (EU's) Emission Trading Scheme (ETS)—a trading program that covers CO_2 emissions from roughly 12,000 sources across 27 countries—fell drastically when it became evident that policymakers had overallocated emission allowances.

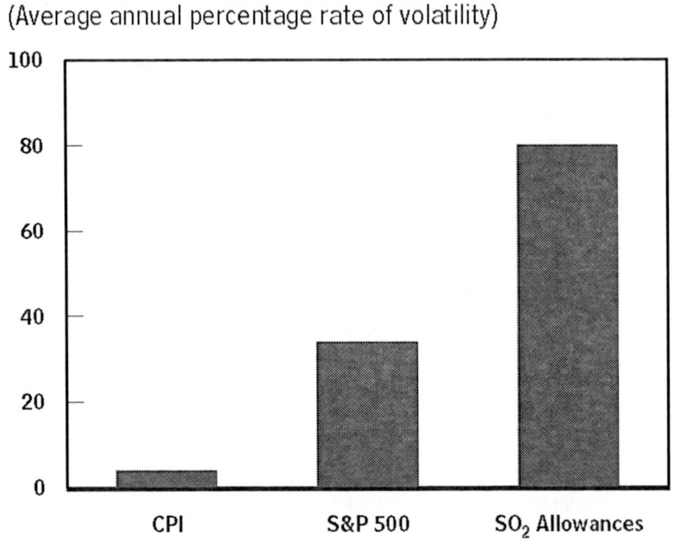

Source:Congressional Budget Office based on William D. Nordhaus, "To Tax or Not to Tax: Alternative Approaches to Slowing Global Warming," *Review of Environmental Economics and Policy*, vol. 1, no. 1 (Winter 2007), pp. 26–44.

Note:Volatility is calculated as the annualized absolute logarithmic month-to-month change in the consumer price index (CPI), the stock price index for the Standard & Poor's 500 (S&P 500), and the price of sulphur dioxide (SO_2) allowances under the U.S. Acid Rain Program.

Figure 1.3. Volatility of SO_2 Allowance Prices and Selected Other Prices, 1995 to 2006

Price volatility could be particularly problematic with CO_2 allowances because fossil fuels play such an important role in the U.S. economy. They accounted for 85 percent of the energy consumed in the United States in 2006. CO_2 allowance prices could affect energy prices, inflation rates, and the value of imports and exports. Volatile allowance prices could have disruptive effects on markets for energy and energy-intensive goods and services and make investment planning difficult.[21] The smoother price path offered by a CO_2 tax would better enable firms to plan for investments in capital equipment that would reduce CO_2 emissions (for example, by increasing efficiency or using low-carbon fuels) and could provide a more certain price signal for firms considering investing in the development of new emission-reduction technologies.

Conditions under Which a Cap Could Be More Efficient Than a Tax

To compare the net benefits of a tax and a cap, researchers must estimate the marginal benefit of reducing a ton of CO_2 emissions. The efficiency advantage of a tax over a cap, however, does not depend on any particular measure of that benefit or even on the ability to place a monetary value on it. Rather, the advantage of a tax stems from the cumulative nature of climate change and from the fact that a tax is better able to reduce emissions over time without imposing potentially disruptive and unnecessarily expensive annual limits on emissions.

The relative advantages of a tax and a cap could change over time, however. One area of growing concern is that the buildup of greenhouse gases in the atmosphere could cause the global temperature to reach a critical level after which further growth in emissions could trigger a rapid increase in damage.[22] The existence of such a threshold could alter the assumption that the marginal benefit of reducing emissions would be relatively constant and could make a cap more efficient than a tax.

Although concerns about thresholds exist, analysts who have tried to define more precisely the conditions that would cause a cap to be more efficient than a tax have concluded that those conditions are quite narrow and unlikely to apply in the near term. Specifically, scientists would need to have fairly precise knowledge about the location of an emissions threshold, and the threshold would have to be sufficiently close that the government would want to make very large cuts in emissions each year to avoid crossing it.[23] If, instead, policymakers wanted to stabilize the concentration of greenhouse gases in the atmosphere after a period of several decades (at a level that would be expected to prevent the global temperature from rising to a trigger level), there could be considerable leeway about when the reductions took place. A tax would provide flexibility in the timing of emission reductions by encouraging companies to cut emissions more in years when the cost of doing so was low and cutting less when the cost was high. A rigid cap would not provide that flexibility over time.

A fundamental change in the cost of reducing emissions could also reverse the efficiency rankings of a tax and a cap. A cap could become more efficient than a tax if a new technology provided the opportunity to make extremely large cuts in emissions at a low and fairly constant cost, rather than at a rising marginal cost.

Other Efficiency Implications of a Tax or a Cap

Besides the efficiency advantages described earlier, a tax on CO_2 emissions could offer another advantage. By generating a significant amount of revenue, it would give the government a chance to use the revenue in a way that would lower the cost to the economy of curbing emissions. For example, studies have found that the economy-wide cost of reducing emissions could be more than twice as high if the reduction was achieved through a cap-and-trade program (with allowances allocated for free) than if it was achieved through a CO_2 tax (with the revenue used to reduce existing taxes that discourage economic activity, such as taxes on capital, labor, or income).[24] A cap-and-trade program could offer a similar opportunity, but only if the government chose to sell the allowances rather than give them away.

If the government elected to tax CO_2 emissions or sell allowances for them, it could opt to use some of the revenue to achieve other aims as well. One goal could be to offset the adverse financial impact of a CO_2 tax or cap on low-income households, who would bear a disproportionate burden (relative to their income) from the higher energy prices that the policy would trigger. In addition, lawmakers could compensate workers in carbon-intensive sectors (such as the coal industry) who might lose their jobs because of the policy.[25]

Flexible Cap Designs

A cap on CO_2 emissions could achieve some of the efficiency advantages of a tax while maintaining the basic structure of a cap-and-trade program by incorporating various design features to make the cap more flexible. Such policies would allow the cap to be exceeded or altered depending on economic circumstances that affect the cost of reducing emissions.

A Ceiling or Floor on Allowance Prices

Combining an emissions cap with a ceiling on the price of allowances—or safety valve—could offer some of the advantages of a tax.[26] Under that approach, if the cost of cutting emissions (as indicated by the price of allowances) rose to the safety-valve level, the government would issue an unlimited number of allowances at that price, thus allowing emissions to exceed the cap. However, unlike a tax, a cap with a safety valve would not give firms and households an incentive to make additional emission cuts if the cost of doing so was lower than anticipated.

In the illustrative example described above, if a cap limiting CO_2 emissions to 6.3 billion metric tons in 2017 (437 million tons below the baseline level for that year) included a safety-valve price of $15 per metric ton of carbon, it would produce the same outcome as a tax of $15 per ton if the cost of meeting the cap was higher than expected (see Figure 1-1 on page 3). In that case, both the tax and the cap/safety valve policy would allow higher emissions than an inflexible cap and would limit the cost of reductions to $15 per ton. Conversely, if the cost of meeting the cap was lower than expected, the cap/safety valve would produce the same outcome as an inflexible cap. The lower-than-expected costs would cause net benefits to be higher than anticipated, but not as high as they would be with a tax.

Under some circumstances, a cap with a safety valve could offer roughly half of the efficiency gains of a tax over a rigid cap. That situation would be most likely to occur if the safety-valve price was set at the amount of the most efficient tax (assumed to be $15 per ton of CO_2 in this example) and the cap was set at the level of the most efficient inflexible cap (estimated to be 6.3 billion metric tons, on the basis of an assumed marginal benefit of $15 per ton of CO_2 and the quantity of emission reductions that would result from that price).[27] In that case, the net benefits of the cap/safety valve policy would fall roughly halfway between those of a cap and a tax (see Figure 1-2 on page 6).

If the safety-valve price was kept at the level of the most efficient tax but the cap was tightened, then the cap/ safety valve policy would function more like a tax and would become even more efficient (see Figure 1-4). Specifically, the amount of emission reductions would increasingly depend on the cost limit specified by the safety-valve price rather than on the quantity limit specified by the cap. At the extreme, a cap of zero emissions with a safety-valve price of $15 per ton of CO_2 would provide the same incentives as a tax of $15 per ton. The cap of zero emissions would not prohibit emissions, but companies would have to purchase an allowance from the government at the safety-valve price for each ton of CO_2 they emitted. (Adding banking or a price floor to a cap-and-trade program with a safety valve offers another way to capture more of the efficiency advantages that could result from an appropriately designed tax. That option is discussed later in this chapter.)

In the other direction, if the cap in the cap/safety valve approach remained at the level of the most efficient inflexible cap but the safety-valve price rose above the level of the most efficient tax, then the cap/safety valve policy would function more like an inflexible cap and would become less efficient. In that case, the amount of emission reductions would be more likely to be determined by the cap than by the safety-valve price. At the extreme, if the safety-valve price was raised high enough that the safety valve would not be triggered, the policy would be equivalent to not having a safety valve, and the net benefits would be the same as those of an inflexible cap.

A recent criticism of a safety valve is that it could unintentionally reduce firms' incentives to replace carbon-intensive capital equipment and to develop new technologies for lowering CO_2 emissions.[28] Either taxing or capping emissions would set a price on them. Researchers generally conclude that the most efficient price for CO_2 emissions would be relatively low in the near term but would rise substantially over time. Expectations of higher future prices would give companies an incentive to gradually replace their stock of physical capital associated with carbon-intensive energy use (such as coal-fired generators or inefficient heating systems) and to invest in researching and developing new technologies that would reduce emissions (such as improvements in solar power, wind power, or energy efficiency).[29] The higher that future allowance prices were expected to rise, the greater that incentive would be. Including a safety valve in a cap-and-trade program, however, would lower expectations about future prices by ensuring that the price of allowances would not rise above the safety-valve level, although it could fall below. In other words, the fact that the range of potential future prices would be truncated at the high end by the safety valve but not at the low end would reduce the expected price.[30] As a result, the safety valve could have the unintended effect of inducing less capital-stock turnover and less investment in research and development (R&D) than would occur under an inflexible cap or a tax.[31]

That problem could be addressed by adding a floor on allowance prices.[32] Enforcing a minimum price for allowances could be fairly straightforward if the government chose to sell a significant share of the allowances rather than give them to affected businesses for free. If allowances were auctioned, policymakers could specify a reserve auction price and restrict the supply of allowances to maintain that price. In combination, a reserve price and a safety valve could define a band of acceptable clearing prices for the allowance market in a cap-and-trade system and could stabilize price expectations. Thus, that combined policy could capture much of the efficiency advantage offered by a tax on emissions (see Figure 1-2 on page 6).[33]

Enforcing a minimum price would be considerably more difficult if nearly all of the allowances were given away for free. In that case, the government could attempt to enforce a minimum price only by reducing the supply of allowances—for example, it could buy back allowances from firms or decrease the value of allowances so that each allowance would permit less than one ton of emissions. Determining when such actions should be undertaken would require the government to make judgments about current and future allowance prices (for example, distinguishing short-term dips from long-term trends). To the extent that those judgments were incorrect, the adjustments to the supply of allowances might undercorrect or overcorrect the allowance price. Further, some analysts are concerned that identifying a trigger price at which policymakers would alter the cap could actually promote price volatility. For example, firms might resist buying allowances once the price began to approach or exceed the trigger point, waiting for policymakers to loosen the cap. But once the demand for allowances dropped, the price would begin to fall and the possibility of intervention would diminish. As a result, purchases (and prices) would once again begin to increase.[34]

Alternatively, increasing the stringency of the cap, while holding the safety valve constant, would reduce the potential problem of underinvestment in R&D and insufficient capital-stock turnover. As noted above, the safety valve would become increasingly likely to determine the quantity of emission reductions and the price of allowances. It would also keep the amount of reductions from falling below the efficient level when the cost of cutting emissions was low. Another option that could help address the underinvestment problem would be to allow emitters to bank allowances for future use.

Banking and Borrowing Allowances

Banking and borrowing would give firms the opportunity to move allowances—and the emissions that correspond to them—between time periods. Each emission allowance would be valid for a specific year or alternative compliance period. (A 2017 allowance, for example, would allow the company that held it to emit one ton of emissions in that year.) With banking, a company could reduce its emissions below the amount it would be permitted to emit on the basis of its allowance holdings for a given year, thereby using fewer allowances in that year, and could bank the extra allowances to use in a future year.[35] With borrowing, by contrast, a firm could exceed its permitted level of emissions in one year by borrowing from its allocation of allowances for a future year.

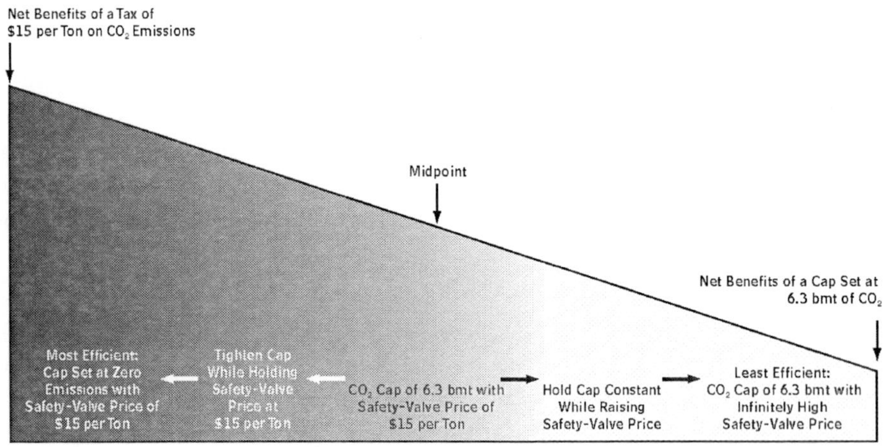

Source: Congressional Budget Office based on information from Richard G. Newell and William A. Pizer, *Indexed Regulation,* Discussion Paper 06-32 (Washington, D.C.: Resources for the Future, June 2006).

Note: CO_2=carbon dioxide; bmt = billion metric tons.

Figure 1.4. Illustrative Range of Net Benefits for a Cap With a Safety Valve Compared With a Tax or an Inflexible Cap on CO_2 Emissions

Emitters would want to bank allowances in years when they thought the price of allowances was low relative to that of future years (for example, because of a mild winter or a period of slow economic activity, or because they believed that tighter caps in the future would lead to higher allowance prices). Conversely, companies would want to borrow allowances in years when they thought the price of allowances was high relative to that of future years (for example, because they expected a new, low-cost technology for reducing emissions to become available later).

Banking Allowances

Banking provisions could improve the efficiency of a cap-and-trade program, regardless of whether the program included a safety valve. While a safety valve could prevent the price of allowances from climbing too high, banking could help prevent the price from falling lower than policymakers would like. Firms would have an incentive to bank allowances in a given year if the cost of making *additional* emission reductions in the current year—that is, reductions in excess of the aggregate amount that firms need for compliance in that year—was less than the expected present value of the cost of reducing emissions or buying allowances in the future. By providing firms with an incentive to save their own allowances—or purchase additional allowances for saving—banking would boost the demand for, and the price of, allowances in years in which that price was relatively low.[36]

The combination of banking and a safety valve could help keep the marginal cost of emission reductions in line with their anticipated benefits under some conditions. For example, such a policy could be effective in preventing relatively short-term lows in allowance prices, but it would be less effective in boosting the price of allowances if the cost of reducing emissions turned out to be significantly lower than anticipated in both the near term and the long term—because of the introduction of a new technology, for instance.[37] In that case, the market price for allowances could stay well below the safety-valve price (that is, below the expected marginal benefits), and the policy would motivate too few emission reductions. As discussed above, policymakers could help ensure that the safety valve would be triggered by setting the cap relatively tightly in comparison with the safety-valve price (see Figure 1-4).

Provided that the safety valve was expected to be eliminated at some point, combining a safety valve with banking provisions could create an incentive for firms to purchase very large amounts of allowances through the safety-valve mechanism and bank them for use once the safety valve was removed.[38] That strategy could prevent a sharp increase in the price of allowances once the safety valve was removed, but it could also mean that the cap would not be met for several years after the removal. The potential for such an outcome would be greatest if the safety valve was holding the price of allowances well below the actual cost of meeting the cap. For example, suppose that firms were allowed to buy allowances through the safety valve in 2020 for $20 but that the safety valve was expected to be removed in 2021 and that, in its absence, the price of allowances required to actually meet the 2021 cap was anticipated to be $40. In that case, firms would have an incentive to purchase very large quantities of allowances through the safety valve in 2020 and use those allowances once the safety valve was removed.

The large excess supply of allowances purchased through the safety valve would prevent the steep jump in allowance prices that would have occurred if firms had not been allowed to bank allowances, but it would also mean that the annual cap in 2021—and for a period of time thereafter—would not be met, even though the safety valve was no longer in place. If policymakers wished to reduce the potential for a multiyear delay in attaining the cap after the safety valve was removed, they could require firms to use allowances purchased at the safety-valve price in the year in which they were purchased.[39]

In addition, policymakers could choose to sell safety-valve allowances through an auction—rather than at a given price—and specify a reserve price for the auction that would increase as greater quantities of allowances were sold in any given year. For example, policymakers could choose to auction blocks of allowances, with increasing reserve prices, just prior to each year's compliance deadline. The reserve price could be $22 for the first block, for instance, $24 for the second block, and so on. Such a strategy could prevent the price of allowances from jumping up once the safety valve was removed while limiting firms' incentives to bank a large supply of allowances for use in future years.[40]

Borrowing Allowances

Including either borrowing provisions or a safety valve in a cap-and-trade program could help prevent spikes in the price of allowances; however, a safety valve could offer greater efficiency advantages. Borrowing would help bring down the price of allowances in a given year only if the price in that year was high relative to prices anticipated in future years. For example, if the price of allowances was $30 in 2010 and was expected to be $15 in 2015, then a firm would have an incentive to borrow 2015 allowances for use in 2010. If, however, the price was expected to be $45 in 2015, no such incentive would exist. Thus, borrowing could help avoid a price spike but would not necessarily keep the cost of emission reductions from exceeding their expected benefits. A safety valve, in contrast, could prevent the cost of emission reductions from exceeding estimates of the benefit of those reductions.

Allowing firms to make one-for-one trades between current and future allowances (and, correspondingly, between current and future emissions) would provide them with too much incentive to defer emission reductions to the future. Because firms discount future costs relative to current costs, they would have an incentive to engage in borrowing (and, thus, defer the cost of reducing emissions) simply to delay the cost of reducing emissions. The potential for excessive borrowing could be avoided if the government discounted borrowed allowances at the rate that companies use to discount future costs.[41] That rate will generally vary from firm to firm; however, policymakers would need to choose a single discount rate. Some researchers suggest that the government could use a discount rate equal to the industry average interest rate used to finance mediumterm capital expenditures.[42] In addition, policymakers could choose to limit the amount of borrowed allowances that companies might use for compliance in any given period or the length of time over which borrowing might occur.

Policymakers could attempt to enforce a ceiling on the price of allowances (for example, keeping it roughly in line with the expected benefits of reducing emissions) by altering the terms under which firms could borrow allowances.[43] Reducing restrictions on borrowing or lowering the rate at which borrowed allowances were discounted could increase the supply of borrowed allowances and thus reduce allowance prices in the near term. As described above, such a strategy could only be effective if firms anticipated that the price of emission reductions in the future would be low (in present-value terms) relative to the current price of allowances. (If that was not the case, firms would not have an incentive to borrow, even under the revised terms.) As a result, altering the terms under which firms might borrow allowances would be more effective in dealing with relatively short-term price spikes than with a situation in which policymakers had underestimated the cost of compliance—in both the near term and in the future—when they set the level of the cap.

Using such a strategy to enforce a limit on the price of allowances would require policymakers to have relatively accurate information about both the current and future prices of allowances. To the extent that those estimates were wrong, the changes that policymakers made to borrowing terms could over or undercorrect the price. For example, if policymakers reduced restrictions on borrowing in order to lower the current price of allowances, but market conditions changed, the increased supply of allowances could cause their price to drop more than policymakers had intended. Alternatively, the increased availability of allowances might fail to reduce the current price as much as policymakers had anticipated.

Circuit Breaker

Some analysts have suggested that an emissions cap that declined at a preset rate and that included a "circuit breaker" would offer economic advantages relative to an inflexible cap and perhaps relative to a cap with a safety valve as well. The circuit breaker would freeze the cap if the price of an allowance exceeded a specified level.[44]

Provided that the circuit breaker price was set at an efficient level (that is, the level that reflected the best available information on costs and benefits), a cap-and-trade program with a circuit breaker could be more efficient than a rigid cap. Specifically, it would offer some economic relief if the cost of meeting the declining cap was higher than the anticipated marginal benefits. Unlike a safety valve, however, a circuit breaker would not set an upper limit on the cost of reducing emissions. Once the circuit breaker was triggered and the cap stopped declining, the allowance price could continue to increase (albeit by not as much as if the circuit breaker was absent). In fact, continued price increases would be likely because meeting a constant cap would become increasingly costly as the economy grew. Thus, assuming that the circuit breaker price was set equal to the expected marginal benefits of reducing emissions, the allowance price (and the cost of achieving additional emission reductions) would be likely to rise above those expected benefits.

2. Implementation Considerations for Different Policy Designs

In addition to the efficiency trade-offs highlighted in the previous chapter, policymakers may wish to weigh other aspects of alternative policies, including the likelihood that the policy could be easily implemented. The following discussion examines the relative ease of implementing a carbon dioxide tax, an inflexible cap, and the flexible cap designs discussed in the previous chapter. It is not meant to provide a comprehensive examination of the challenges associated with implementing individual policies but rather to highlight implementation considerations that would vary across policies.

A Carbon Dioxide Tax versus an Inflexible Carbon Dioxide Cap

Successfully implementing either a CO_2 tax or an inflexible cap would entail several similar requirements. Under an upstream design, suppliers of fossil fuels (such as coal producers, petroleum refiners, and natural gas processors) would be required to pay a tax—or hold an allowance—for each ton of carbon that was contained in the fuel they sold (and, thus, would be emitted in the form of CO_2 when the fuel was burned). In that case, firms would need to report their sales data and the carbon content of the fuels they sold so that regulators could determine each firm's tax or allowance requirement. Regulators would need to have methods of verifying the accuracy of the reported data. In that way, they could detect under-payments of taxes or excessive emissions and impose adequate, consistent, and predictable penalties.

Further, regulators would need to have a method of ensuring that all fuels that should be subject to the regulatory requirements were covered by the policy. That would entail accounting for fossil fuels that did not pass through a domestic mine mouth (less than 0.5 percent of all coal consumed in the United States), a domestic petroleum refinery (approximately 1 percent of the petroleum produced or imported into the United States), or a natural gas processing plant (approximately 22 percent of the natural gas consumed in the United States).[45] Finally, regulators would need to be able to accurately identify fossil fuels that were not combusted and, therefore, should be exempt from the tax or allowance requirement, such as petroleum that was used in producing plastics or tires. On the basis of information from the Energy Information Administration, such a system would entail regulating roughly 150 oil refineries, 1,460 coal mines, and 530 natural gas processing plants.[46]

Moving the point of regulation downstream—to users of fossil fuels—could be more difficult to implement in some sectors. For the power sector, such a change would be relatively simple to make because large power producers subject to the Acid Rain Program are required under the Clean Air Act to have equipment in place that continuously monitors CO_2 emissions. (See the appendix for a description of the Acid Rain Program.) Outside the power sector, however, a downstream system could impose significant implementation challenges. The number of entities that would need to be regulated would grow, and identifying their emissions would initially be difficult. In fact, inaccurate data about the baseline emissions of downstream industries (such as cement, iron, and steel plants) in the European Union's trading program for CO_2 emissions caused regulators to issue more allowances to those industries than they intended to under the first phase of the program. That overallocation contributed to large price swings at the end of the first year of reporting.[47]

Similar data problems occurred with the start-up of the Acid Rain Program. It took the Environmental Protection Agency two rounds of data review with industry (through public notice and comment) over the course of two years to sort out anomalies in the data used for determining generating units' initial allocations (based on energy data reported to the Department of Energy). For example, EPA made adjustments for electricity generating units that had significant outages during the period that was used to determine the initial allocations. Because EPA had a much longer time to implement the Acid Rain Program than the EU had to implement the initial phase of its program for carbon dioxide, EPA was able to review and revise the data before allocating the allowances. And because the anomalies were discovered before the initial allocations were made and the trading program was operational, the revisions did not lead to price swings in the allowance market.[48]

Another implementation consideration is whether allowances should be grandfathered, or given away for free, on the basis of previously existing circumstances. A cap-and-trade program in which allowances were not grandfathered could have substantially lower start-up costs (because it would avoid the lengthy process of determining the basis for grandfathering) than a cap-and-trade program in which allowances were grandfathered. A tax would have significantly lower start-up costs than a cap-and-trade program with grandfathering provided that policymakers did not decide to grant exemptions based on historical production or emissions data. Further, implementing a tax would not require the government to set up a process for auctioning allowances.

The cost of implementing an upstream carbon tax is likely to be less than that of a cap-and-trade program (regardless of how allowances were initially allocated) because the tax could build upon an existing infrastructure. For example, coal producers already pay an excise tax (which is used to fund the Black Lung Trust Fund) as do producers and importers of petroleum (to fund the Oil Spill Trust Fund). A CO_2 tax based on the sales of coal or petroleum would be an additional excise tax and could, presumably, be implemented at a relatively modest incremental cost. While natural gas is not subject to a federal excise tax, many natural gas processors are subject to a corporate income tax.

In contrast, implementing an upstream cap-and-trade program would probably require a new administrative infrastructure. However, based on EPA's experience with the Acid Rain Program, the cost of administering such a program could be relatively modest. Regulators would need to take the following steps:

- Set up an allowance account for each regulated unit and for other nonregulated entities that might wish to trade allowances (such as brokers),
- Record information on allowance allocations for each regulated unit,
- Review submitted allowance transfers to make sure that they have all necessary information and meet the regulatory requirements,
- Record transfers into and out of each account, and
- Notify both participants in a transfer when the transfer was recorded.[49]

EPA estimates that it spends approximately $1.5 million annually to operate its Allowance Tracking and Allowance Transfer Systems for the Acid Rain Program.[50] That program maintains accounts for regulated power generators (who must comply with the cap-and-trade program) as well as for other traders. On the basis of the most recent data, a little more than half of the accounts (roughly 1,200) are held by regulated generators, and the remainder (915) are general accounts.[51] In 2005, nearly 5,700 private allowance transfers (moving roughly 20 million allowances) were recorded in EPA's Allowance Tracking System.[52]

Flexible Cap Designs

As discussed in the previous chapter, including features in a cap-and-trade program that would make it more responsive to annual variations in the cost of reducing emissions could improve its efficiency. In some cases, those features could be relatively easy to incorporate:

- Implementing a safety valve (a ceiling on the price of emission allowances) could be relatively straight-forward. The government could offer an unlimited amount of allowances at the safety-valve price.
- Implementing banking provisions (in which firms could save allowances from one period to use in a future period) could also be straightforward. Banking has already been successfully implemented in several existing cap-and-trade systems. For example, emitters in the Acid Rain Program may bank allowances for an unlimited amount of

time, and some countries participating in the European Union's Emission Trading Scheme allowed banking in the first phase of the program (2005 to 2007). For implementation purposes, borrowing (in which firms use allowances designated for a future period in the current period) would be similar to banking.

Other features of a cap-and-trade program could be more difficult to implement. A circuit breaker, which would freeze an otherwise declining cap once the price of allowances rose to a predetermined circuit-breaker price (and would keep the cap at that level until the allowance price fell back below the circuit-breaker price) could pose more significant implementation challenges. In order to determine when to trigger the circuit breaker, policymakers would need accurate information on allowance prices. They would also need to decide how sensitive the trigger would be. For example, would the circuit breaker be triggered if any single allowance was traded at a price above the circuit-breaker price? Or would it be based on a price index? If so, would the chosen price indicator have to remain above the circuit-breaker price for a given amount of time? Making such determinations could be difficult, for several reasons:

- Allowances for CO_2 could be traded in "over the counter" transactions between an individual buyer and seller (possibly through a broker)—as is the case for the sulfur dioxide (SO_2) allowances that are traded under the Acid Rain Program. In such transactions, the parties involved are not required to report the price at which the commodity is traded. In the case of SO_2, most brokers voluntarily report prices, and several publications report prices or publish indexes—but those prices are not verified.[53]
- Traders could have an incentive to provide inaccurate information about prices. For example, sellers of allowances could inflate their price information to convince buyers that they need to pay more for their allowances. Likewise, regulated entities might wish to have the price of allowances appear high enough to trigger the circuit breaker so as to prevent the CO_2 cap from becoming more stringent.
- Prices could fluctuate widely over time. Determining whether a change in price represented a temporary spike or a more permanent shift in underlying market conditions would be difficult.
- There could be several different prices for allowances at any given point, and those prices would vary during a year. As a result,

determining when the circuit breaker should be triggered could be difficult: Policy-makers would need to decide which allowance price the circuit breaker would be based on and how long that price would have to be above the specified level to cause the cap to stop declining. Policymakers would need to make similar decisions in order to determine when the circuit breaker should no longer be in effect—and the cap should once again begin to decline.

3. International Consistency Considerations for Different Policy Designs

Carbon dioxide is a global pollutant: A ton of emissions from any point on the globe (at a given time) would have the same effect on the atmospheric concentration of CO_2 and, therefore, would result in the same amount of damage. As a consequence, the most cost-effective method of achieving a given atmospheric concentration of CO_2 would be to undertake the lowest-cost emission reductions, regardless of where those opportunities were located. Achieving that goal would require that major emitting countries coordinate their policies to create a consistent economic incentive to reduce emissions. Choices that U.S. policymakers might make could affect the feasibility of creating such an incentive.

As in the previous chapter, this discussion is not meant to provide a comprehensive examination of the challenges in coordinating policies with other countries but rather to highlight how the ability to achieve that goal might vary across the policy designs. For example, effective government institutions and legal systems in each country would be necessary to successfully implement any type of multinational tax or cap-and-trade program and, therefore, would not give one policy a comparative advantage over another. Further, this discussion focuses primarily on the efficiency implications of creating a consistent economic incentive to reduce emissions in major emitting countries and touches only briefly on the potential equity issues associated with achieving that goal.

A Carbon Dioxide Tax versus an Inflexible Carbon Dioxide Cap

Major emitting countries could achieve a uniform price on CO_2 by agreeing to implement the same tax on emissions (that is, to harmonize their countries' policies). Alternatively, each country could establish a national cap-and-trade program and agree to link their programs by permitting allowance trading across borders. In that case, competitive forces would lead to a single allowance price.

Harmonizing a U.S. Tax on CO_2 with Policies in Other Countries

A direct method of achieving a uniform price on CO_2 across multiple countries would be for each country to adopt the same tax. For example, each country might agree upon a specific tax rate, such as $15 per metric ton of CO_2. (That tax rate was used as an illustrative example in Chapter 1.) A uniform tax rate would ensure an equal level of incentive to reduce emissions in participating countries only if the following conditions were met:

- Participating countries had equally effective monitoring and enforcement provisions. Less effective monitoring, lower penalties, or less rigorous enforcement in any given country would reduce the economic incentive provided by its tax and would be equivalent to reducing the country's tax rate.[54]
- Participating countries agreed on similar tax exemptions or other special provisions. For example, if one country provided an exemption for the steel industry, that industry would have a reduced incentive to cut its emissions and would have a competitive advantage over steel industries in other countries with the same tax but no exemption.
- Participating countries implemented the tax at the same point in the carbon supply chain or made special provisions for differences in the point of implementation. For example, a country with an upstream tax on fossil fuel suppliers would need to exempt fossil fuels that were sold to a country with a downstream tax on fossil fuel users in order to avoid double-taxing emissions.

Alternatively, the United States could choose to implement a CO_2 tax set at a rate to be consistent with the price of CO_2 in an outside cap-and-trade system, such as the European Union's Emission Trading Scheme (see the appendix). Such a tax could only roughly approximate the allowance price because allowance prices are difficult to predict and can fluctuate widely over time.[55] Further, attempts to harmonize the CO_2 tax rate in the United States with the allowance price in an outside trading program would have to take into account differences in the point of implementation. For example, if the United States adopted an upstream tax, it would need to exempt any fossil fuels that were sold to countries participating in the EU's ETS, because that system regulates emissions at the point of combustion.

Linking a U.S. Cap-and-Trade Program with Outside Cap-and-Trade Programs

Linking the cap-and-trade programs of multiple countries to achieve a uniform price of CO_2 would involve the same complications associated with harmonizing tax rates. As with a tax, participating countries would need similar monitoring of emissions, tracking of allowance transactions, penalties for noncompliance, and enforcement provisions. In contrast with a harmonized tax, lax monitoring or enforcement in one country would undermine the effectiveness of the policy not only in that country but in other participating countries as well. The country with lax enforcement could become a supplier of fraudulent allowances (ones that did not correspond to actual reductions), diminishing the environmental integrity of the entire trading system.[56] Further, the systems that track and transfer allowances in different countries (referred to as "registries" in the EU) would need to be able to communicate with each other.[57] Finally, as with a harmonized tax, each country's cap-and-trade program would need to cover similar sources of emissions, and provisions would need to be made to avoid double-charging (or not charging for) emissions if countries applied their caps at different points in the carbon supply chain.

Linking cap-and-trade programs would also entail additional challenges beyond those associated with harmonizing a tax on CO_2. Linking would change the price of allowances in each participating country, which would alter gains and losses and could create incentives for strategic behavior. A country with a relatively high allowance price (because of a more stringent cap, for example, or a greater dependence on high-carbon fuels) would experience a price decrease as a result of linking. In contrast, a country with a relatively low price before linking would see an increase. Those price changes would have several effects that countries would need to consider:

- The change in the price of allowances would alter the gains and losses experienced by companies that, before linking, had been net buyers or net sellers of allowances. For example, if the United States experienced an increase in the price of allowances as a result of linking, U.S. firms that had been net sellers could benefit, whereas net buyers could be worse off.

- In addition to altering the gains and losses experienced by individual firms, linking would create net flows of allowances—and flows of resulting revenues—into, or out of, countries. Countries could have an incentive to choose their caps strategically so as to take advantage of those potential flows. For example, a country might try to choose a less stringent cap so that it could become a net supplier of allowances.[58]
- A change in the price of allowances as a result of linking could alter the incentive of domestic producers to invest in new technologies—such as energy efficiency improvements or alternative fuels—that would reduce CO_2 emissions.

Linking would remove a country's ability to determine the terms of regulation for its own businesses. For example, if a country that did not allow its firms to borrow future allowances for current use was to link with a country that did, firms in both countries would have access to borrowed allowances. In a similar manner, the use of other flexible design features—such as banking, offsets, and a safety valve (discussed in the next section)—would be available to all firms in a linked system should any one country allow its firms to comply in those ways.

Flexible Cap Designs

Design features that could make a U.S. cap-and-trade program more efficient than an inflexible cap could make other countries more or less willing to link their cap-and-trade program with a U.S. program. The following discussion examines linkage considerations associated with efficiency-improving design features discussed in the previous chapter: a safety valve, a price floor, banking and borrowing provisions, and a circuit breaker. It does not address other design features that could influence whether a country decides to link its trading system with a U.S. system. Those features might include U.S. decisions about how to allocate allowances to domestic sources or decisions about whether to allow sources to comply by using offsets such as biological sequestration (capturing carbon for long-term storage in trees or soil), geological sequestration (capturing carbon and storing it in the ocean or in the earth), and projects designed to reduce emissions in developing countries.[59]

Including a safety valve in a U.S. cap-and-trade program could limit the likelihood that countries participating in a system with an inflexible cap, such as the EU's ETS, would be willing to link with a U.S. program. That reluctance could stem from two concerns. First, if the EU agreed to link with a U.S. program, it would no longer be able to maintain a rigid cap because EU sources would have access to allowances at the safety-valve price.[60] In addition, the U.S. government could receive significant revenue by selling allowances to EU firms.

Including a safety valve in a U.S. cap-and-trade program could limit the likelihood that countries participating in a system with an inflexible cap, such as the EU's ETS, would be willing to link with a U.S. program. That reluctance could stem from two concerns. First, if the EU agreed to link with a U.S. program, it would no longer be able to maintain a rigid cap because EU sources would have access to allowances at the safety-valve price.[60] In addition, the U.S. government could receive significant revenue by selling allowances to EU firms.

Linking a U.S. cap-and-trade program with trading programs in other countries could limit the ability of the U.S. government to set a floor on the price of allowances, even if it chose to sell a significant fraction of domestic allowances in an auction. Linking could greatly expand the size of the allowance market, which, in turn, would lessen the government's ability to affect their price by withholding allowances from the domestic auction.

As with a safety valve, if one country in a multinational cap-and-trade program chose to allow its emitters to bank or borrow allowances, then those options could become available to all emitters within the system, regardless of their location. For example, if firms in one country were allowed to bank allowances (for example, in 2010), those additional allowances would be available through the allowance trading market to firms in all countries in the linked trading system in a future year (for example, in 2015). Banking could be problematic if some countries had binding targets that had to be met within a given period, however. That concern has caused EU countries to prevent emitters from banking allowances from the first phase (2005 to 2007) of its ETS for use in the second phase, which has binding targets for the 2008– 2012 period.[61]

One of the challenges in designing an efficient global approach to reducing CO_2 emissions is how to include developing countries. Those countries have contributed a small fraction of global emissions in the past, but they are expected to become major contributors in the future. Some researchers suggest that linking a system of fixed cap-and-trade programs could offer an opportunity to equalize the marginal cost of emission reductions among participating countries while allowing for a differentiated level of effort among countries (that is, some countries could be required to make larger emission reductions than others) based on fairness or other criteria.[62] Other researchers suggest that the revenue generated by taxing CO_2 emissions—or by selling allowances—in developed countries could be used to fund emission reductions in developing countries.[63]

APPENDIX A. CURRENT AND PROPOSED CAP-AND-TRADE PROGRAMS IN THE UNITED STATES AND EUROPE

The concept of distributing tradable pollution rights—what this paper refers to as emission allowances—first appeared in the academic literature in 1968.[64] Trading programs can be attractive alternatives to more traditional approaches that mandate specific pollution limits for all sources. A primary advantage of trading programs is that they can lower the costs of achieving a given environmental goal by giving participants some flexibility about where and how reductions are made.

Trading programs have been used for various purposes in the United States, such as to decrease the amount of lead in gasoline, to reduce discharges into rivers and reservoirs, and to lower emissions of two air pollutants—sulfur dioxide (SO_2) and nitrous oxide (NO_x). The trading programs for SO_2 and NO_x provide the most relevant comparison for a trading program for carbon dioxide (CO_2) emissions. Trading programs to reduce such emissions have been proposed in the United States and are in effect in Europe.

U.S. Programs for Sulfur Dioxide and Nitrous Oxide

The United States has two major emissions cap-and-trade programs that cover multiple states.[65] The Acid Rain Program is a nationwide program that caps SO_2 emissions from large electric power units. The program took effect in 1995 and was expanded to cover additional units in 2000. It currently covers about 3,000 generating units at more than 700 power plants. The initial free allocation of allowances was based on each unit's fuel input in the mid-1980s, multiplied by an emissions performance standard. Units were allocated 30 years' worth of allowances, and those allowances could be banked indefinitely, meaning that an allowance for a ton of emissions in any given year could be used in that year or in any future year.

The Acid Rain Program is run by the Environmental Protection Agency (EPA) and is widely viewed as being very successful, bringing about large reductions in SO_2 emissions for lower-than-expected costs. Banking provisions contributed to the program's cost-effectiveness, but the free allocation of allowances did not. The method of allocating allowances would have a substantial impact on the distribution of policy costs but, in general, would not affect the overall cost of achieving a cap. (An exception is that free allocations to regulated utilities could increase the cost of achieving a cap by preventing price increases that are essential for triggering cost-effective emission reductions.) Further, giving allowances to firms, as opposed to selling them, could preclude the government from using the proceeds from selling allowances to reduce existing taxes that dampen economic activity.[66]

The NO_x Budget Trading Program is a multistate trading program that caps nitrous oxide emissions from large industrial boilers and electricity generating units in 19 states, the District of Columbia, and portions of two additional states. That program, which originally encom-passed nine northeastern states in the late 1990s, is a partnership between the federal government and state governments. States have responsibility for allocating emission allowances; the EPA implements an emissions and allowance registry, verifies emissions data, runs the trading program, and reconciles emissions and allowances (to determine compliance) at the end of each year. As under the SO_2 trading program, the NO_x allowances are bankable.

In addition to those multistate programs, the South Coast Air Quality Management District, a local air pollution agency in southern California, has operated the Regional Clean Air Incentives Market (RECLAIM) since 1994. That program caps SO_2 and NO_x emissions from various sectors (including the power sector and some industrial sectors). Unlike the Acid Rain and NO_x Budget Trading programs, no banking is allowed under RECLAIM because of concern that banking would lead to unacceptably high emissions in a future year. The lack of banking is thought to have contributed to a severe price spike for NO_x emission rights in California in 2000 (see Chapter 1).

U.S. and European Programs for Greenhouse Gase

No mandatory cap-and-trade programs for greenhouse-gas emissions such as carbon dioxide currently exist in the United States, but state-level efforts to develop them are under way. For example, 10 states—Connecticut, Delaware, Maine, Maryland, Massachusetts, New Hampshire, New Jersey, New York, Rhode Island, and Vermont—are developing a multistate cap-and-trade program covering greenhouse-gas emissions, the Regional Greenhouse Gas Initiative (RGGI). RGGI will begin capping emissions in 2009 and will initially cover CO_2 emissions from power plants in participating states. In the future, RGGI may be extended to include other sources of CO_2 emissions and other greenhouse gases.[67]

Further, the state of California is actively considering the feasibility of implementing a cap-and-trade program for CO_2 emissions. In September 2006, California enacted legislation that directs the California Air Resources Board (CARB) to establish a comprehensive program that would reduce the state's greenhouse-gas emissions to 1990 levels by 2020.[68] The legislation does not specifically require the use of a market-based system, such as a cap-and-trade program, but instructs CARB to consider other proposed or existing trading programs, including RGGI.

The largest cap-and-trade program for CO_2 emissions at present is the European Union's Emission Trading Scheme (ETS). The initial phase of the ETS—the warm-up phase—went into effect in 2005 and continued through 2007. The second phase, which is in effect from 2008 through 2012, coincides with the initial phase of the Kyoto Protocol. The ETS currently covers carbon dioxide emissions from roughly 12,000 sources across the 27 countries of the European Union. Sources of covered emissions include factories that produce iron and steel, cement, glass and ceramics, pulp and paper, electric power, and petroleum products. Other greenhouse gases and other sectors, such as aviation, may be added in the future. Allowances valued at \$23 billion and covering more than 1 billion metric tons of emissions were traded in the EU's ETS in 2006.

The warm-up phase of the ETS provides several lessons for avoiding potential problems in the future. For example, observers of the program note that member states had insufficient historic emissions data for some participating installations. As a result, some member states based their allocations of allowances on estimates rather than actual emissions. The resulting inaccuracies led to caps that were less stringent than anticipated, and the market price for allowances dropped significantly when that overallocation became apparent.[69]

End Notes

[1] For more information about U.S. cap-and-trade programs for sulfur dioxide and nitrous oxide and about the European Union's program for carbon dioxide, see the appendix.

[2] See, for example, William A. Pizer, "Combining Price and Quantity Controls to Mitigate Global Climate Change," *Journal of Public Economics,* vol. 85 (2002), pp. 409–434; Michael Hoel and Larry Karp, "Taxes and Quotas for a Stock Pollutant with Multiplicative Uncertainty," *Journal of Public Economics,* vol. 82 (2001), pp. 91–114; and Richard G. Newell and William A. Pizer, "Regulating Stock Externalities Under Uncertainty," *Journal of Environmental Economics and Management,* vol. 45 (2002), pp. 416–432.

[3] For more information about the implications of placing a cap upstream or downstream, see Congressional Budget Office, *An Evaluation of Cap-and-Trade Programs for Reducing U.S. Carbon Emissions* (June 2001).

[4] For example, coal producers pay an excise tax that is used to fund the Black Lung Trust Fund, and petroleum producers and importers pay an excise tax that finances the Oil Spill Trust Fund.

[5] Department of Energy, Energy Information Administration, *International Energy Annual 2005* (updated September 18, 2007), Table H.1co2, available at *www.eia.doe.gov/* iea/ carbon. html.

[6] This point was made by Robert N. Stavins in "Linking Tradable Permit Systems: Opportunities, Challenges, and Implications" (paper presented at the 7th International Emissions Trading Association's Forum on the State of the Greenhouse Gas Market, Washington, D.C., September 27, 2007).

[7] See Joseph E. Aldy, Peter R. Orszag, and Joseph E. Stiglitz, "Climate Change: An Agenda for Global Collective Action" (paper prepared for the Pew Center on Global Climate Change's work-shop "The Timing of Climate Change Policies," Washington, D.C., October 11–12, 2001); and Joseph E. Aldy, Scott Barrett, and Robert N. Stavins, *13+1: A Comparison of Global Climate Change Policy Architectures*, Discussion Paper 03-26 (Washington, D.C.: Resources for the Future, August 2003).

[8] See William A. Pizer, "Combining Price and Quantity Controls to Mitigate Global Climate Change," *Journal of Public Economics,* vol. 85 (2002), p. 416.

[9] For a more detailed discussion, see Congressional Budget Office, *Uncertainty in Analyzing Climate Change: Policy Implications* (January 2005).

[10] For more discussion of policy choices in the face of catastrophic costs, see Cass R. Sunstein, *Worst-Case Scenarios* (Cambridge, Mass.: Harvard University Press, 2007).

[11] The stringency of emission-reduction policies is sometimes discussed in terms of carbon and sometimes in terms of CO_2. Estimated costs or benefits that appear in dollars per ton of CO_2 can easily be translated into dollars per ton of carbon by multiplying by the ratio of the molecular weight of CO_2 to the molecular weight of carbon (44/12, or 3.67). Thus, a tax of $15 per ton of CO_2 translates into a tax of $55 per ton of carbon. Conversely, costs and benefits that are stated in terms of dollars per ton of carbon can be converted into dollars per ton of CO_2 by dividing by 3.67.

[12] For a description of how CBO calculated the emission reductions that would result from a given tax, or the price of allowances that would result from a given cap, see Mark Lasky, *The Economic Cost of Reducing Emissions of Greenhouse Gases: A Survey of Economic Models,* CBO Technical Paper 2003-03 (May 2003).

[13] The cost of reducing emissions in any given year is incurred in that year, while the benefits accrue over a period of decades or centuries. Thus, comparing the costs and benefits of emission reductions involves discounting the value of future benefits to the current year. This illustrative example assumes that the benefits of reducing a ton of emissions have a present value of $15. As a result, reducing emissions by 437 million metric tons would produce benefits of $6.55 billion. The cost of achieving those reductions would be $3.07 billion, according to Lasky, *The Economic Cost of Reducing Emissions of Greenhouse Gases.*

[14] The cost changes considered in this example correspond to two separate doublings of the price sensitivity parameter. Thus, the cost of cutting emissions by 437 million metric tons doubles from $7.50 to $15 per metric ton and then from $15 to $30 per metric ton.

[15] For a more detailed discussion of the uncertainty about the costs and benefits of emission reductions, see Congressional Budget Office, *Uncertainty in Analyzing Climate Change: Policy Implications* (January 2005), pp. 30–31.

[16] See Pizer, "Combining Price and Quantity Controls to Mitigate Global Climate Change." That paper considered a worldwide tax or cap on carbon emissions. In analyzing the sensitivity of his results to how long the policies are assumed to remain in place, the author assumed that the damage from climate change would rise rapidly once a certain temperature increase had occurred (in other words, that the damage function was sharply kinked). In that case, a cap would yield larger net benefits than a tax. However, the difference ($600 billion) would be small compared with the net benefits offered by either policy (roughly $34 trillion). Thus, under a sharply kinked damage function, the paramount concern would be to make drastic cuts in emissions, and the choice of policy tool would be relatively unimportant.

[17] Michael Hoel and Larry Karp, "Taxes and Quotas for a Stock Pollutant with Multiplicative Uncertainty," *Journal of Public Economics,* vol. 82 (2001), pp. 91–114. Only under the assumptions of very great damage from climate change and a large initial stock of allowances do those authors conclude that a cap would be more efficient.

[18] See Richard G. Newell and William A. Pizer, "Regulating Stock Externalities Under Uncertainty," *Journal of Environmental Economics and Management,* vol. 45 (2002), pp. 416–432.

[19] William D. Nordhaus, "To Tax or Not to Tax: Alternative Approaches to Slowing Global Warming," *Review of Environmental Economics and Policy,* vol. 1, no. 1 (Winter 2007), pp. 26–44.

[20] See A. Denny Ellerman, Paul L. Jaskow, and David Harrison Jr., *Emissions Trading in the U.S.: Experience, Lessons, and Considerations for Greenhouse Gases* (Arlington, Va.: Pew Center on Global Climate Change, May 2003), pp. 24–25, available at www.pew climate.org/global-warming-in-depth/all_reports/emissions_ trading. Some observers argue that the lack of banking provisions contributed to the price spikes. Such spikes could have been prevented by the inclusion of a safety valve as well. (Those design features are discussed later in this chapter.)

[21] Nordhaus, "To Tax or Not to Tax," pp. 37–39.

[22] See National Research Council, *Abrupt Climate Change: Inevitable Surprises* (Washington, D.C.: National Academy Press, 2002), pp. 13–14; R.B. Alley and others, "Abrupt Climate Change," *Science*, vol. 229 (March 28, 2003), pp. 2005–2010; and Congressional Budget Office, *Uncertainty in Analyzing Climate Change*, Box 2-1, pp. 10–11.

[23] See William A. Pizer, *Climate Change Catastrophes*, Discussion Paper 03-31 (Washington, D.C.: Resources for the Future, May 2003).

[24] See Congressional Budget Office, *Trade-Offs in Allocating Allowances for CO_2 Emissions* (April 25, 2007).

[25] Ibid.

[26] That feature is included in a cap-and-trade proposal (S. 1766) introduced by Senator Bingaman on July 11, 2007.

[27] As determined in Lasky, *The Economic Cost of Reducing Emissions of Greenhouse Gases.*

[28] See Dallas Burtraw and Karen Palmer, "Dynamic Adjustment to Incentive Based Policy to Improve Efficiency and Performance" (draft, Resources for the Future, Washington, D.C., November 30, 2006).

[29] The amount of investment in research and development under either a tax or a cap-and-trade program could be less than the amount that would be best for society because such investment may generate "spillover benefits" to society that do not translate into profits for the firm doing the investing. For a discussion of that issue, see Congressional Budget Office, *Evaluating the Role of Prices and R&D in Reducing Carbon Dioxide Emissions* (September 2006).

[30] For example, suppose policymakers set a cap on emissions in 2020, and observers generally agreed that there was a 25 percent chance that the allowance price necessary to meet the cap would be $25, a 50 percent chance that it would be $50, and a 25 percent chance that it would be $75. With no safety valve, the expected allowance price would be $50 [that is, (0.25 x $25) + (0.50 x $50) + (0.25 x $75)]. If, however, policymakers set a safety valve at $50, the expected allowance price would fall to $43.75 [(0.25 x $25) + (0.75 x $50)].

[31] That effect is not reflected in Figure 1-4.

[32] Burtraw and Palmer, "Dynamic Adjustment to Incentive Based Policy to Improve Efficiency and Performance."

[33] If both the price floor and the safety valve were set at the expected marginal benefit of emission reductions, the combined policy would be analogous to a tax.

[34] See Ian W.H. Parry and William A. Pizer, "Emissions Trading Versus CO_2 Taxes Versus Standards," in Raymond J. Kopp and William A. Pizer, eds., *Assessing U.S. Climate Policy Options: A Report Summarizing the Work at RFF as Part of the InterIndustry U.S. Climate Policy Forum* (Washington, D.C.: Resources for the Future, November 2007), pp. 83–84.

[35] Uncertainty about the existence of a cap-and-trade program in the future would undermine incentives for banking.

[36] See Henry D. Jacoby and A. Denny Ellerman, "The Safety Valvemate Policy Forum (Washington, D.C.: Resources for the Future, and Climate Policy," *Energy Policy*, vol. 32, no. 4 (March 2004),November 2007), pp. 83–84. pp. 481–491.

[37] For a discussion of this point, see Burtraw and Palmer, "Dynamic Adjustment to Incentive Based Policy to Improve Efficiency and Performance."

[38] This observation was made by William A. Pizer of Resources for the Future in a personal communication to the Congressional Budget Office.

[39] That requirement would reduce, but not eliminate, the delay. Firms would be able to comply in 2020 by using safety-valve allowances and then banking 2020 allowances that they had obtained by other means (such as receiving for free, making reductions, or purchasing from other firms).

[40] This suggestion was offered by William A. Pizer of Resources for the Future.

[41] If each allowance let firms emit one ton of CO_2, a borrowed allowance could permit a firm to emit less than one ton, with the amount of the reduction depending on the discount rate that policymakers chose and the number of years in the future from which the reduction was borrowed. Alternatively, policymakers could allow firms to emit one ton of emissions for each borrowed allowance but could require that they reduce emissions by more than one ton when they pay back the allowance loan.

[42] See Catherine Kling and Jonathan Rubin, "Bankable Permits for the Control of Environmental Pollution," *Journal of Public Economics,* vol. 64, no. 1 (April 1997), p. 112.

[43] For example, that feature is included in the cap-and-trade proposal (S. 2191) introduced by Senators Lieberman and Warner on October 18, 2007.

[44] See the statement of Joel Bluestein before the Subcommittee on Clean Air, Climate Change, and Nuclear Safety of the Senate Committee on Environment and Public Works, May 8, 2003.

[45] Based on information provided to the Congressional Budget Office by the Environmental Protection Agency's Clean Air Markets Division (July 5, 2007).

[46] See Department of Energy, Energy Information Administration, "EIA-820, Annual Refinery Report" as of January 1, 2006, and "EIA-816, Monthly Natural Gas Liquids Report," both available at www.eia.doe.gov/oss/forms.html, and EIA's 2005 Coal Production Data Files, available at www.eia.doe.gov/cneaf/coal/page/ database.html.

[47] Once this fact was revealed, prices of allowances fell by more than 75 percent. See the statement of Jill Duggan, Head of International Emissions Trading, U.K. Department for Environment, Food, and Rural Affairs, *EU Cap-and-Trade Programme,* before the House Committee on Energy and Commerce (March 29, 2007), p. 3.

[48] Based on information provided to the Congressional Budget Office by the Environmental Protection Agency's Clean Air Markets Division (July 5, 2007) as well as Joseph Kruger and William A. Pizer, *The EU Emissions Trading Directive: Opportunities and Potential Pitfalls* (Washington, D.C.: Resources for the Future, April 2004), pp. 14–15.

[49] Those steps are based on EPA's responsibilities for operating both the Allowance Tracking System and the Allowance Transfer System for sulfur dioxide trading under the Acid Rain Program. See Environmental Protection Agency, *Information Collection Request Renewal for the Acid Rain Program Under the Clean Air Act Amendments Title IV* (July 26, 2006), pp. 31–32.

[50] Ibid., p. 32.

[51] Based on information provided to the Congressional Budget Office by the Environmental Protection Agency's Clean Air Markets Division (July 5, 2007).

[52] See Environmental Protection Agency, *Acid Rain 2005 Progress Report,* EPA-430-R-06-15 (October 2006), p. 9.

[53] Based on information provided to the Congressional Budget Office by the Environmental Protection Agency's Clean Air Markets Division (July 5, 2007).

[54] In addition, countries would need to be prevented from changing their tax codes in order to neutralize the effect of the carbon tax. See Joseph E. Aldy, Scott Barrett, and Robert N. Stavins, *13+1: A Comparison of Global Climate Change Policy Architectures,* Discussion Paper 03-26 (Washington, D.C.: Resources for the Future, August 2003), p. 13. For a discussion of some potential methods of inducing international compliance—such as using economic sanctions, social sanctions, "carrots," or other indirect incentives—see Joseph E. Aldy, Peter R. Orszag, and Joseph E. Stiglitz, "Climate Change: An Agenda for Global Collective Action" (paper prepared for the Pew Center on Global Climate Change's workshop "The Timing of Climate Change Policies," Washington, D.C., October 11–12, 2001).

[55] Prices depend on numerous factors, including the stringency of the cap, available technologies, supply and demand conditions in energy markets, and monitoring and enforcement provisions.

[56] See Richard Baron and Stephen Bygrave, *Towards International Emissions Trading: Design Implications for Linkages* (Paris: Organisation for Economic CoOperation and Development and International Energy Agency, October 2002), p. 21.

[57] See Joseph A. Kruger and William A. Pizer, "Greenhouse Gas Trading in Europe: The New Grand Policy Experiment," *Environment*, vol. 46, no. 8 (October 2004), p. 15.

[58] See Jane Ellis and Dennis Tirpak, *Linking GHG Emission Trading Schemes and Markets* (Paris: Organisation for Economic Co-Operation and Development and International Energy Agency, October 2006), p. 24; and Erik Haites, "Harmonisation Between National and International Tradeable Permit Schemes: CATEP Synthesis Paper," in *Greenhouse Gas Emissions Trading and Project-Based Mechanisms* (Paris: Organisation for Economic Co-operation and Development's Global Forum on Sustainable Development, Emissions Trading CATEP Country Forum, March 17–18, 2003), p. 107.

[59] For a discussion of the implications of those design features for linking, see Ellis and Tirpak, *Linking GHG Emission Trading Schemes and Markets;* and Kruger and Pizer, "Greenhouse Gas Trading in Europe."

[60] See Ellis and Tirpak, *Linking GHG Emission Trading Schemes and Markets,* p. 26. Even if EU firms were prohibited from purchasing U.S. allowances through the safety-valve mechanism, U.S. entities could serve as intermediaries: They could purchase safety-valve allowances for their own use, freeing up other allowances to sell to firms in the European Union.

[61] Ibid., p. 23.

[62] This point was made by Robert N. Stavins in "Linking Tradable Permit Systems: Opportunities, Challenges, and Implications" (paper presented at the 7th International Emissions Trading Association's Forum on the State of the Greenhouse Gas Market, Washington, D.C., September 27, 2007).

[63] See Aldy, Orszag, and Stiglitz, "Climate Change: An Agenda for Global Collective Action"; and Aldy, Barrett, and Stavins, *13+1: A Comparison of Global Climate Change Policy Architectures.*

[64] See J.H. Dales, *Pollution, Property, and Prices* (Toronto: University of Toronto Press, 1968). Also see David W. Montgomery, "Markets in Licenses and Efficient Pollution Control Programs," *Jounal of Economic Theory*, vol. 5 (1972); and Tom H. Tietenberg, *Emissions Trading: An Exercise in Reforming Pollution Policy* (Washington, D.C.: Resources for the Future, 1985).

[65] For a summary of these programs, see Joseph A. Kruger and William A. Pizer, "Greenhouse Gas Trading in Europe: The New Grand Policy Experiment," *Environment,* vol. 46, no. 8 (October 2004), p. 14. For more detailed descriptions of the multistate trading programs for sulfur dioxide and nitrous oxide, see the "Clean Air Markets" section of the Environmental Protection Agency's Web site, available at www.epa.gov/airmarkets (with links to "Acid Rain Program" and "NOx Trading Programs"). For a detailed description of those programs as well as the Regional Clean Air Incentives Market program, see A. Denny Ellerman, Paul L. Joskow, and David Harrison Jr., *Emissions Trading in the U.S.: Experience, Lessons, and Considerations for Greenhouse Gases* (Arlington, Va.: Pew Center on Global Climate Change, May 2003), *www.pewclimate.org/global-warming-in-depth/all_reports/* emissions_ trading.

[66] For a discussion of the distributional and efficiency aspects of alternatives for allocating allowances, see Congressional Budget Office, *Trade-Offs in Allocating Allowances for CO_2 Emissions* (April 2007).

[67] For more information, see the Regional Greenhouse Gas Initiative's Web site at www.rggi.org.

[68] See Jonathan Ramseur, *Climate Change: Action by States to Address Greenhouse Gas Emissions,* CRS Report for Congress RL33812 (Congressional Research Service, January 18, 2007).

[69] See Kruger and Pizer, "Greenhouse Gas Trading in Europe"; and Senate Committee on Energy and Natural Resources, "Full Committee Roundtable: European Union's Emissions Trading Scheme" (March 26, 2007), available at http://energy.senate.gov/public/ index.cfm?FuseAction=Hearings.Hearing&Hearing_ID=1615. In addition, the United Kingdom initiated a voluntary emissions-trading system in 2002. Thirty-three organizations adopted emission-reduction targets to reduce their emissions against 1998–2000 levels. That trading scheme ended in December 2006. See "UK Emissions Trading Scheme" at www.defra.gov.uk/ environment/climatechange/trading/uk/index.htm.

In: Policy Option Issues for CO2 Emissions ISBN: 978-1-60741-381-3
Editors: Nikolaus Vogler

Chapter 4

ROLE OF PRICES AND R&D IN REDUCING CARBON DIOXIDE EMISSIONS

Congressional Budget Office

1. SUMMARY AND INTRODUCTION

Several important human activities—most notably the worldwide burning of coal, oil, and natural gas—are gradually increasing the concentrations of carbon dioxide and other greenhouse gases in the atmosphere and, in the view of many climate scientists, are gradually warming the global climate. That warming, and any long-term damage that might result from it, could be reduced by restraining the growth of greenhouse gas emissions and ultimately limiting them to a level that stabilized atmospheric concentrations.

The magnitude of warming and the damages that might result are highly uncertain, in part because they depend on the amount of emissions that will occur both now and in the future, how the global climate system will respond to rising concentrations of greenhouse gases in the atmosphere, and how changes in climate will affect the health of human and natural systems. The costs of restraining emissions are also highly uncertain, in part because they will depend on the development of new technologies.[1] From an economic point of view, the challenge to policy-makers is to implement policies that balance the uncertain costs of restraining emissions against the benefits of avoiding uncertain damages from global warming or that minimize the cost of achieving a target level of concentrations or level of annual emissions.

Researchers have studied the relative efficacy—as well as the appropriate timing—of various policies that might discourage emissions of carbon dioxide (referred to as carbon emissions in the rest of this paper), which makes up the vast majority of greenhouse gases, and restrain the growth of its atmospheric concentration. This paper presents qualitative findings from that research, which are largely independent of any particular estimate of the costs or benefits of reducing emissions. The paper's conclusions are summarized below.

Policies for Reducing Carbon Emissions

The possibility of climate change involves two distinct "market failures" that prevent unregulated markets from achieving the appropriate balance between fossil fuel use and changes in climate. One market failure involves the external effects of emissions from the combustion of fossil fuels—that is, the costs that are imposed on society by the use of fossil fuels but that are not reflected in the prices paid for them. The other market failure is a general underinvestment in research and development (R&D) that occurs because investments in innovation may yield "spillover" benefits to society that do not translate into profits for the innovating firm. The first market failure yields inefficiently high use of fossil fuels; the second yields inefficiently low R&D.

Because there are two separate market failures, an efficient response is likely to involve two separate types of policies:

- One type of policy would reduce carbon emissions by increasing the costs of emitting carbon, both in the near term and in the future, to reflect the damages that those emissions are expected to cause.
- The other type of policy would increase federal support for R&D on various technologies that could help restrain the growth of carbon emissions and would create spillover benefits.

Policymakers could increase the cost of emitting carbon by setting a price on those emissions. That could be accomplished by taxing fossil fuels in proportion to their carbon content (which is released when the fuels are burned) or by establishing a "cap-and-trade" program under which policymakers would set an overall cap on emissions but allow fossil fuel suppliers to trade rights (called allowances) to those limited emissions. Either a tax or a cap-and-trade program would cause the prices of goods and services to rise to reflect the amount of carbon emitted as a result of their consumption. To the extent that a carbon tax or allowance price reflected the present value of expected damages,[2] such policies would encourage users of fossil fuels to account for the costs they impose on others through their emissions of greenhouse gases.

Researchers generally conclude that the appropriate price for carbon would be relatively low in the near term but would rise substantially over time, resulting in relatively modest reductions in emissions in the near term followed by larger reductions in the future. Phasing in price increases would allow firms to gradually replace their stock of physical capital associated with energy use and to gain experience in using new technologies that emit less carbon. Firms would have an incentive to invest in developing new technologies on the basis of their expectations about future prices for emissions.

Federal support could be provided for the research and development of technologies that would lead to lower emissions. Such technologies could include improvements in energy efficiency; advances in low or zero-emissions technologies (such as nuclear, wind, or solar power); and development of sequestration technologies, which capture and store carbon for long periods. Federal support would probably be most cost-effective if it went toward basic research on technologies that are in the early stages of development. Such research is more likely to be underfunded in the absence of government support because it is more likely to create knowledge that is beneficial to other firms but that does not generate profits for the firm conducting the research.

The Interaction and Timing of Policies

Pricing and R&D policies are neither mutually exclusive nor entirely independent—both could be implemented simultaneously, and each would tend to enhance the other. Pricing policies would tend to encourage the use of existing carbon-reducing technologies as well as provide incentives for firms to develop new ones; federal funding of R&D would augment private efforts; and successful R&D investments would reduce the price required to achieve a given level of reductions in emissions.

Neither policy alone is likely to be as effective as a strategy involving both policies. Relying exclusively on R&D funding in the near term, for example, does not appear likely to be consistent with the goal of balancing costs and benefits or the goal of minimizing the costs of meeting an emissions reduction target. At any point in time, there is a cost continuum for emissions reductions, ranging from low-cost to high-cost opportunities. Unless R&D efforts virtually eliminated the value of near-term reductions in emissions (an outcome that appears unlikely given reasonable assumptions about the payoff of R&D efforts), waiting to begin initial pricing (to encourage low-cost reductions) would increase the overall cost of reducing emissions in the long run.

Near-term reductions in emissions achieved with existing technologies could be valuable even if fundamentally new energy technologies would be needed to prevent the buildup of greenhouse gases in the atmosphere from reaching a point that triggered a rapid increase in damages. Near-term reductions could take advantage of low-cost opportunities to avoid adding to the stock of gases in the atmosphere and could allow additional time for new technologies to be developed and put in place. That additional time could prove quite valuable, given that R&D efforts are highly uncertain and that the process of putting new energy systems in place could be slow and costly.

Determining the appropriate mix of policies to address climate change is complicated by the fact that future policies would be layered on a complex mix of current and past policies, all of which affect today's use of fossil fuels and their alternatives as well as the amount of R&D. The analyses reviewed in this paper typically do not account for existing policies or for the administrative costs of implementing a carbon-pricing program or of initiating a larger (and perhaps redesigned) R&D program for carbon-reducing technologies. However, the qualitative conclusion reached in those analyses—that costs would be minimized by a combination of gradually increasing emissions prices coupled with subsidies for R&D—is not likely to be affected by such considerations.[3]

A Global Concern

The causes and consequences of climate change are global, and reductions in U.S. emissions alone would be unlikely to have a significant impact. Cost-effective mitigation policies would require coordinated international efforts and would involve overcoming institutional barriers to the diffusion of new technologies in developing countries, such as India and China. If a domestic carbon-pricing program significantly increased the prices of U.S.-produced goods—and was not matched by efforts to reduce emissions in other countries—it could cause carbon-intensive industries to relocate to countries without similar restrictions, diminishing the environmental benefits of a domestic program.

However, successful domestic R&D efforts, whether funded by the public or private sector, could lower the costs of reducing carbon emissions in other countries as well as within the United States. Some new technologies, such as those that yielded improvements in energy efficiency, might be deployed without additional incentives. Other innovations, such as sequestration technologies or alternative energy technologies that reduce carbon emissions but cost more than their fossil-fuel-based alternatives, would be unlikely to be deployed without financial incentives to reduce carbon emissions.

2. Policies for Reducing Carbon Emissions

Rising emissions of carbon dioxide from the combustion of fossil fuels, and the resulting increases in the concentration of carbon in the atmosphere, raise concerns about the prospect of climate change and the costs that could be imposed on society. If policymakers chose to take action to reduce the level of carbon emissions, they would face two distinct "market failures" that would prevent the free market from settling on the amount of carbon emissions that would be best for society. In general, distinct market failures are best addressed by separate policy instruments.

One market failure arises because the combustion of fossil fuels may create external costs that are borne by society as a whole—particularly by future generations—but are not reflected in the prices that people pay for those fuels, or for the services that the fuels provide. As a result, firms, households, governments, and other organizations would be likely to consume more fossil fuels—and emit more carbon—than would be best for society. Setting a price on carbon emissions that reflected their external costs would correct that market failure.

The other market failure arises because the research and development of new technologies for reducing carbon emissions are likely to result in spillover benefits. Such benefits are realized by other firms as a result of the innovating firm's R&D effort but do not generate profits for the innovating firm. Thus, firms tend to engage in less R&D on carbon-reducing technologies than would be best for society. Increasing public funding for the development of new carbon-reducing technologies would correct that failure, as long as the spillover benefits of such activities exceeded their costs.

Current policies in the United States both encourage and discourage firms and households from using fossil fuels, but no existing policy provides systematic incentives for users to account for the potential costs of carbon emissions. Current U.S. policies also provide subsidies for R&D that at least partially compensate for spillover benefits; determining whether those subsidies are sufficient, however, is beyond the scope of this paper.

Pricing Carbon to Address the External Costs of Emissions

The costs of carbon emissions could be built into the price of consuming fossil fuels directly by taxing such fuels or indirectly by allowing the market to set a price on carbon emissions through a cap-and-trade program. A tax could be imposed on fossil fuel producers and importers on the basis of the carbon content of their fuels, a so-called upstream tax. Alternatively, an upstream cap-and-trade program would require producers or importers to pay for the right (called an allowance) to sell goods and services that ultimately led to carbon emissions. The cost of the tax or the allowance, like any other cost, would be reflected in the price producers charged for goods and services that resulted in carbon emissions (see Box 2.1). Either approach could be designed to provide firms with an incentive to capture and sequester carbon emissions, thus reducing net emissions into the atmosphere, as well as to reduce their initial emissions.

The Effects of Pricing

Establishing a price for carbon emissions would cause the costs of using fossil fuels to rise to reflect their carbon content, with the largest percentage increase for coal, followed by oil and then natural gas.[4] Higher fossil fuel costs and the resulting increases in the prices of services, such as electricity, would provide incentives for emitters to make changes in their behavior or operations that would conserve energy (such as driving less and turning down thermostats); to invest in more-efficient appliances and equipment; and to make greater use of renewable fuels to heat water and power vehicles. The carbon price would also provide an incentive for electricity producers to decrease the carbon intensity of the electricity that they produced. That decrease could be accomplished by switching to fossil fuels with lower carbon contents (for example, from coal to natural gas), by generating electricity from nuclear power or renewable fuels, or from converting coal to gas and capturing and sequestering the carbon.[5] The magnitude of the incentive created for those activities would depend on the price of the carbon.[6]

Box 2.1. How Emissions Taxes and Cap-and-Trade Policies Would Work

Policymakers could take several approaches to limiting carbon dioxide emissions. One approach—a tax on those emissions—would raise the cost of emitting carbon dioxide, thereby encouraging households and firms to cut their emissions as long as the cost of doing so was less than the tax. That approach would set an upper limit on the cost of individual reductions in emissions (at the level of the tax) but would not ensure that any particular emissions target was met.

A second approach—a cap-and-trade program— would set an overall limit on the level of carbon dioxide emissions but leave the decisions about where and how the necessary reductions should be made to households and firms. Under that approach, policy-makers would establish an overall cap on emissions but allow regulated firms to trade rights to those emissions, called allowances. That trading would permit firms that could reduce their emissions most cheaply to sell some of their allowances to firms that faced higher costs to reduce their emissions. Such an approach would limit the overall level of emissions but would not place any explicit limit on the cost of individual reductions.

Under a third, hybrid approach, policymakers would set an overall cap on total emissions, but they would also establish an upper limit on the price of allowances, referred to as a "safety valve" price. If the price of allowances rose to the safety-valve price, the government would sell as many allowances as was necessary to maintain that price. Thus, if the safety valve was triggered, the actual level of emissions would exceed the cap. The cap would be met only if the price of allowances never rose above the safety-valve price.

Research has shown that, given the uncertainty associated with the costs and benefits of carbon-reduction policies, a tax on carbon dioxide emissions (or a cap-and-trade program with a safety valve) would result in significantly higher expected net benefits than a cap-and-trade program with a fixed cap.[1] A tax would motivate people to limit their emissions up to the point at which the costs of doing so were equal to the tax. If actual costs were less than, or greater than, anticipated, people would limit their emissions more than, or less than, policymakers projected. However, emissions would be reduced up to the point at which the cost of doing so was equal to the expected benefits (provided that the tax was set at the efficient level). In contrast, a strict cap on emissions could result in actual costs that were far greater (or less than) expected and, therefore, exceeded (or fell below) the expected benefits.

Either a tax on carbon dioxide emissions or a cap-and-trade program could be designed to provide incentives for firms to sequester carbon. For example, a tax credit for sequestration could be implemented in conjunction with a tax on carbon dioxide emissions. Likewise, firms could be permitted to reduce their allowance requirements by sequestering carbon.

[1] For a more detailed discussion of this point, see Congressional Budget Office, *Limiting Carbon Dioxide Emissions: Prices Versus Caps*, CBO Issue Brief (March 15, 2005).

Carbon pricing would encourage cost-effective reductions in emissions at any given point in time because it would provide firms, households, governments, and other organizations with a uniform incentive to under-take the broad array of activities that would reduce emissions. A worldwide pricing policy would be more efficient than one that was limited to the United States because carbon is a global pollutant. A uniform world-wide price for carbon would provide an incentive for emitters that could make reductions at the lowest cost to do so, regardless of national boundaries. Establishing a domestic price for carbon emissions would minimize the cost of achieving a given level of reductions within the United States, but that same quantity of reductions could be obtained at a lower cost under coordinated policies that set an equal price for carbon emissions throughout the world.

Setting a current price for carbon emissions and announcing planned future carbon prices not only would induce firms and households to change their behavior but also would increase their demand for technologies that would reduce emissions. That increase in demand would in turn create incentives for firms to research and develop new methods of improving energy efficiency, producing energy from renewable sources, and sequestering carbon.[7] Moreover, as firms gain more experience with low-carbon technologies, they may learn how to produce them at a lower cost.[8]

Some researchers suggest that future carbon taxes would not provide adequate incentives for R&D if a firm anticipated that policymakers might lower the carbon tax—and thus reduce the firm's return on its R&D investment—once it successfully developed a new technology. That concern could be valid only for a technology that would be expensive to develop and, once developed, could reduce a significant fraction of the world's carbon emissions at a low per-unit price (see Box 2.2).

Setting Prices

A carbon tax or cap, and the resulting increase in fossil fuel prices, would impose costs on the economy; consequently, policymakers may wish to balance those costs against the anticipated reduction in climate-related damages. An efficient perton price for carbon would reflect the damages that are anticipated from a ton of carbon emissions. Damages from today's emissions would occur in the future, and their expected magnitude can only be estimated with a high degree of uncertainty.[9] Provided that the price of carbon was set equal to the current value of the damages anticipated from today's emissions, carbon pricing would give emitters an incentive to under-take reductions when the cost of doing so would be out-weighed by the expected benefit—that is, the avoided damage.

The relationship between near-term and future carbon prices would determine the pattern of reductions in emissions over time. In general, gradually rising prices have been found to be most efficient—that is, most likely to result in a pattern of reductions that would best balance costs and benefits over the long run.[10]

Gradually rising carbon prices would lead to steady reductions in emissions (relative to the baseline trend). Phasing in reductions would allow firms to gradually replace the capital stock that affects carbon emissions—including housing stock, appliances, automobiles, industrial equipment, and power plants—with newer capital that would lead to lower emissions, such as more energy-efficient houses, or power plants that generate electricity from sources that do not emit carbon. In the absence of economic incentives to reduce carbon emissions, such replacements would be unlikely to occur, and a new generation of carbon-intensive capital stock could be put in place—perhaps necessitating premature retirements of that stock in the future.[11]

Box 2-2. Would Taxes on Carbon Dioxide Emissions Spur Research and Development of New Technologies?

A tax on carbon dioxide emissions would create a market for technologies that could reduce those emissions. Emitters would find it worthwhile to purchase those technologies provided that the cost of doing so was less than the cost of paying the tax. Current taxes would create incentives for current reductions in emissions, and expectations of future taxes would provide firms with an incentive to invest in the research and development, or R&D, of new technologies.

Some research has suggested that announced future taxes may not provide an adequate incentive for firms to develop a fundamentally new technology for reducing carbon dioxide emissions. That research is based on the argument that the government may have an incentive to reduce carbon taxes, and correspondingly returns to investors, after costly R&D was undertaken. Specifically, policymakers might announce one set of future taxes—which would reflect the full costs of developing and deploying such a technology—but then reduce those taxes once the technology was developed.[1] If that occurred, emitters would be willing to pay less for the new technology than the initially announced taxes would suggest. If firms anticipated that policymakers would reduce the tax once the new technology was introduced—and that the tax reduction would cause the innovating firm to be unable to recoup some of its R&D costs—then firms would be less willing to undertake the initial R&D investment to develop the technology.[2]

What set of circumstances might lead to the outcome described above? Policymakers might be motivated to reduce taxes on carbon dioxide emissions after a firm successfully developed a new technology in the case of a technology that could reduce a significant fraction of the world's emissions—for example, by producing carbon-free energy—at a low and relatively constant per-unit cost. In that case, the new technology would warrant a decrease in the tax because it would be expected to so greatly reduce the future atmospheric concentration of carbon dioxide that the marginal damage from current and future emissions would be lessened.[3] Maintaining the tax once the new technology was introduced would induce some reductions in emissions that would be unwarranted on the basis of the resources required to achieve them and the now lower benefits (avoided damages) that they would create. An efficient tax reduction would be more likely to prevent the innovating firm from recouping its R&D costs if those costs were large and if the cost of providing an additional unit of the new technology was relatively constant. Thus, the contention that a firm would be dissuaded from developing a new technology out of concern that its introduction would trigger a tax reduction that would prevent the firm from recovering its R&D costs could be valid only for a technology that met two criteria: it would be very expensive to develop, and it would be able to reduce a significant share of the world's carbon dioxide emissions at a low, and relatively constant, per-unit cost.

A technology that fits that case has not been identified, but some so-called silver bullet technologies are being explored. Concern that policymakers might lower a tax on carbon dioxide emissions once such potential technologies were developed could strengthen the case for funding research on them; nevertheless, that concern would be unlikely to pose a general disincentive to other forms of technological innovation that could play an important role in addressing climate change. Finally, even in the case of a silver bullet technology, policymakers could be reluctant to make a tax reduction that eliminated the innovating firm's return on its R&D investment if such an action was expected to discourage future R&D on other carbon-reducing technologies.

Concerns that policymakers might lessen the stringency of the policy once a technological innovation occurred are less likely to apply to a cap-and-trade program that does not include a safety-valve price (see Box 2-1). In that case, technological innovations would encourage policymakers to increase—not decrease—the stringency of the cap (because the cost of achieving any given cap would be reduced), and firms' ability to recoup their R&D expenses would depend primarily on their ability to patent their innovations. (Spillover benefits could still provide a rationale for using public funds to supplement private R&D efforts.) Inclusion of a safety valve could cause a cap-and-trade program to function in a similar manner to a carbon dioxide tax and, thus, could lead to concerns about a tax decrease following a technology innovation under the same set of circumstances described above.

[1] For a discussion of this argument, see W. David Montgomery and Anne E. Smith, "Price, Quantity, and Technology Strategies for Climate Change Policy," in *Human-Induced Climate Change: An Interdisciplinary Assessment* (Boston, Mass.: Cambridge University Press, forthcoming).

[2] As the amount of emissions reductions provided by the new technology increases, the price premium necessary to cover the fixed R&D costs decreases.

[3] Specifically, policymakers would find it efficient to reduce the tax to the point at which the firm's marginal cost of reducing a ton of carbon dioxide with the new technology, which would increase as more of the new technology was deployed, was equal to the marginal damage of emitting an additional ton of carbon dioxide, which would decline as more of the new technology was deployed. The efficient tax would just cover the cost of producing the marginal unit of the new technology, but the firm would earn profits on inframarginal units (which have production costs below the tax). Firms' profits on those inframarginal units may not fully cover the cost of developing the new technology. That situation is most likely to occur when the cost of producing an additional unit of the technology is relatively insensitive to the amount produced and when the R&D costs are very large. Technologies that were not expected to provide enough reductions in emissions to significantly lower the atmospheric concentration of carbon dioxide in the future would increase the efficient amount of emissions reductions but would not be expected to trigger a tax decrease. For a demonstration of the latter point, see Lawrence H. Goulder and Stephen H. Schneider, "Induced Technological Change and the Attractiveness of CO_2 Abatement Policies," *Resource and Energy Economics*, vol. 21 (1999), pp. 211-253.

By providing incentives for relatively low-cost reductions in emissions today and delaying more significant reductions until the future, policymakers would allow time for new technologies to develop. Analysts' opinions about the appropriate timing of reductions in emissions vary, in part, because of different assumptions about how likely it is that technological advances would occur and how costly achieving those advances would be.[12]

Current Policies That Provide Incentives and Disincentives for Fossil Fuel Use

Numerous policies affect the consumption of fossil fuels. Policies that directly or indirectly encourage the use of fossil fuels include federal support for highway construction and various tax provisions that promote the domestic production of oil and natural gas. Policies that discourage the use of fossil fuels include the federal gasoline tax, subsidies for mass transit, and various tax provisions that promote alternative fuels such as ethanol, biogas, and coal synfuels.

Those policies provide conflicting incentives for house-holds and firms to use fossil fuels. Further, even the policies that discourage fossil fuel use do not do so on the basis of the extent to which the fossil fuels contribute to climate change. In contrast, a tax or cap on carbon emissions would provide an incentive for households and firms to take those external costs into account—providing the greatest discouragement for the use of coal, which releases the most carbon when it is burned.[13] Such a tax or cap would encourage a more complete range of activities that could reduce carbon emissions (including improvements in energy efficiency, the use of renewable energy sources, and, depending on the design, sequestration), not just those that are targeted by a specific tax credit.

Subsidizing R&D Efforts to Account for Spillover Benefits

Even if the external costs of carbon were incorporated into prices for fossil fuels, firms might invest less in research and development than would be best from society's point of view. That outcome would be likely if firms' resulting profits were expected to fall short of society's benefits. The extent to which that outcome would occur depends, in part, on the nature of the research, existing patent laws, and existing tax provisions.

The Benefits of Supporting R&D

The development of some technologies to reduce carbon emissions entails basic scientific research. That basic research may create knowledge that is beneficial to society; however, because it may not result in patentable inventions, such research may be underfunded in the absence of federal support. The role of basic research—and the resulting creation of spillover benefits—is likely to be particularly large in developing fundamentally new technologies that are a long way from the marketplace but that could provide large amounts of carbon-free energy, or carbon sequestration, at a low cost. (For example, some researchers have considered the possibility that hydrogen fuel could be manufactured from high-efficiency solar processes or that low-resistivity power lines could distribute solar electricity between continents, time zones, and day/night cycles.)[14]

Other research is likely to be tied to the development of technologies that are much closer to the marketplace (for example, improvements in vehicles' fuel efficiency or in the generation of electricity from wind) and, thus, firms would be more likely to profit from that research. However, even if patents allowed firms to profit from their R&D investments, the level of R&D that would maximize firms' profits could still fall short of the level that would be best for society. Society's benefits would include the value of any new innovations that were inspired by the patented invention but that were not covered by the patent, and the extent to which the benefits created by the patented invention (theoretically measured by an individual's willingness to pay for it) exceeded the price charged for it.

The divergence between private profits from R&D and social gains is not unique to investments designed to reduce carbon emissions. Rather, that divergence is characteristic of R&D investments in general—particularly for basic research and less so for research that is more clearly linked to commercial applications. Measuring the gap between firms' profits from R&D and society's benefits is difficult, but some analysts think that it could be quite large.[15] Should the gap between social benefits and private profits be less than those researchers estimated, the efficiency gains from subsidizing private R&D efforts would be diminished.

The Costs of Supporting R&D

Determining the appropriate amount of resources to devote to subsidizing research efforts is complicated by the fact that research can be costly and outcomes are uncertain. Federal funds devoted to R&D would impose an opportunity cost in that they could not be used for other spending priorities, to lower taxes, or to reduce the deficit. Further, federal funds devoted to inducing more R&D on alternative energy technologies or sequestration could impose a second opportunity cost by "crowding out" R&D in other sectors of the economy. For example, such policies might discourage private R&D spending by coal extraction firms or in other, seemingly unrelated, industries, such as civil engineering and computer software. The latter effect might occur because the policy-induced R&D in one area could bid up the price of key research inputs, such as highly qualified engineers, computer experts, and scientists.[16] Although the magnitude of that effect is unknown, it could be significant. Finally, identifying investments with the highest potential returns can be challenging, although federally funded prizes have been suggested as one method of doing so (see Box 2.3 on pages 12 and 13).

Box 2.3. Using Prizes to Encourage Technological Improvements

The federal government has traditionally funded research and development through the use of tax credits; research grants or contracts to private or academic institutions; or research at federally funded facilities, such as the national laboratories. Such means of funding subsidize the cost of conducting the research rather than provide a monetary award for a successful outcome.

In recent years, analysts and policymakers have considered using prizes as a way to spur technological improvements that would lower the cost of reducing carbon dioxide emissions—such as improvements in energy efficiency, renewable energy sources, or the development of fundamentally new energy sources that emit no carbon dioxide.[1] Prizes have been used to a limited extent to encourage technological developments for hundreds of years. One of the earliest well-known uses of prizes was an award offered by the British government to inventors who designed instruments capable of accurately measuring longitude. In that case, the goal was to reduce the number of ships lost by the Royal Navy. A more-recent example of a federally funded prize is the Defense Advanced Research Projects Agency's "Grand Challenge." That prize, awarded for the development of driverless vehicles, aims to reduce battlefield casualties of U.S. troops. Several prizes have been offered by private entities as well, leading to technological improvements in the areas of aviation and automobiles, among others.

The use of prizes has several advantages over cost-subsidizing approaches:

- Firms that conduct the research bear the risk of failure—that is, they absorb the costs of unsuccessful endeavors. Having researchers, rather than the federal government, bear those costs makes sense for at least two reasons: researchers are likely to have better information on the likelihood of the success of different endeavors, and the research facility itself has the most control over whether it is spending its research dollars effectively.
- Federal award money is directly linked to a successful outcome, with the winning firm unknown at the outset. As a result, federal money could be less influenced by competing goals that policy-makers might have, such as providing employment or research facilities for their states or districts.
- If prizes entail fewer bureaucratic hurdles than the traditional grant or contract process, then they may encourage the participation of smaller firms, those without previous involvement in federally funded research efforts, or both. Some anecdotal evidence indicates that technological break-throughs are more likely to come from more unorthodox entities.

- The prestige associated with winning a competition can offer firms a further incentive to participate.

The use of prizes can have disadvantages as well:

- If the technological improvements sought are risky and expensive, then the prizes necessary to induce them would need to be large.
- If the necessary research is thought to be expensive, then the use of prizes could actually discourage the participation of smaller research entities, which could not cover the costs themselves and may not have access to credit.
- If an award attracted many participants, resources could be wasted as researchers duplicated one another's efforts. While relevant, this limitation is not restricted to prizes. Such duplication can occur as firms attempt to be the first to patent an outcome as well. That problem could be reduced by structuring the competition so that the number of competitors is narrowed over time (through the use of intermediary hurdles) or by requiring competitors to pay to participate.
- Firms would be reluctant to participate if they thought that the government might renege on the prize if attitudes toward the prize's goals changed. To avoid such a possibility, the government could establish a private-sector escrow account or purchase an insurance policy that would guarantee that the funds would be available.

At least four different design considerations may influence the likelihood that offering a prize will lead to technological improvements. First, the technological target must be as specific and measurable as possible, yet broad enough to allow for creative efforts. Capturing the correct balance could be difficult. For example, a general goal of reducing carbon dioxide emissions at a given cost could lead to a wide variety of technological innovations. Some may clearly be more desirable than others, though, because they would increase—or reduce—other forms of pollution (such as nuclear waste). Policymakers may want to take those other costs into account and to specify such considerations beforehand. Focusing climate technology prizes on one research area (such as solar energy) at a time may provide one possible solution.

Second, the amount of the prize must be set sufficiently high to induce participation, but not so high as to provide excessive rewards. Finding the correct balance could be difficult, particularly if the competition is expected to last for an extended period. As described in the main text of this paper, the private rewards from firms' innovations (that is, their profit-making potential) are determined by the market for carbon-reducing technologies, which, in turn, depends on government policies. Should policies that provide incentives for carbon reductions emerge in the future, the private rewards from firms' innovations would be higher than if such policies did not emerge. Ensuring—or restricting—firms' access to patents for their innovations is one way that policy-makers could influence both the amount of the prize and the public availability of the research outcomes. Allowing winning firms to patent their innovations would increase the number of participants, reduce the magnitude of the prize necessary to ensure participation, or both. However, it would also limit the availability of the technology to the public. Further, private firms are likely to have more information about the costs that research endeavors might entail than the policymakers who are responsible for setting the magnitude of the reward. Some researchers have suggested that the latter problem might be resolved by letting researchers bid for the size of the prize they would accept if their efforts were successful.

Third, the conditions for winning would influence the speed and quality of the outcomes. A contest with a cash award could be structured as a tournament (contest) or as a race (first past the post). A tournament, which specifies an objective and a time limit, guarantees an award to the party that has made the most progress toward the specified goals. It encourages participation—parties with substantial uncertainty may enter on the basis of partial insights—but can impose high costs on the government for evaluating many participants' relative progress toward the stated goals. In contrast, a race specifies a well-defined goal (the post to pass) and makes an award only if that goal is achieved. A downside of a race is that it could reward a suboptimal outcome because a competitor who was developing a superior technology may finish second.

Finally, clarity in the rules is essential. Unclear or unenforceable rules can lead to costly litigation and reduce incentives to participate.

[1] For a more detailed discussion of the points made in this box, see the statement of Douglas Holtz-Eakin, Director, Congressional Budget Office, *Economic and Budgetary Issues with Cash Prizes to Achieve NASA's Objectives,* before the Subcommittee on Space and Aeronautics, Committee on Science, U.S. House of Representatives (July 15, 2004); and Richard G. Newell and Nathan E. Wilson, "Technology Prizes for Climate Change Mitigation," Discussion Paper 05-33 (Washington, D.C.: Resources for the Future, 2005).

Studies have reached conflicting conclusions about whether or not fundamentally new technologies for providing carbon-free energy, or for sequestering emissions, would be necessary to prevent the concentration of carbon in the atmosphere from exceeding targeted levels.[17] Such studies, however, provide little guidance for R&D funding decisions because they do not consider either the costs or the benefits of developing the new technologies.

Current Policies That Support R&D

Two tax provisions encourage firms' R&D efforts, thereby increasing the level of R&D they undertake:[18]

- Firms may receive a nonrefundable 20 percent income tax credit for certain research expenditures (such as salaries, time-sharing costs for computers, and contracts). That provision has expired but is expected to be extended.[19]
- Firms may deduct certain research expenditures as a current business expense, even though those expenditures are likely to generate patents with a useful life extending beyond a single tax year.

Those provisions are designed to encourage firms to undertake research on all different forms of technology improvements, not just ones that would reduce carbon emissions. Such tax provisions can increase the profitability of firms' R&D efforts but cannot make up for the lack of a market for a new technology. Thus, in the absence of carbon pricing, those tax provisions would be unlikely to motivate firms to undertake R&D on carbon-reducing technologies that would not be profitable for other reasons (such as rising energy prices). Should carbon pricing be implemented, however, those tax provisions could help boost the amount of investment by closing the gap between private and social gains from the development of new carbon-reducing technologies.

In addition to those tax provisions, the federal government directly funds research that could lead to lower carbon emissions. Examples include research aimed at making advances in producing energy from nuclear fuels, hydrogen, and renewable fuels (such as biomass and solar) as well as efforts to improve the energy efficiency of vehicles, buildings, and industrial processes.

In the absence of carbon pricing, federal support for research on carbon-reducing technologies could help make up for the lack of a profit motive for private firms to undertake such efforts as well as the spillover benefits that private research would have generated. Thus, the efficient amount of federal R&D funding depends, in part, on whether or not policymakers establish policies, such as a carbon tax or cap, that create incentives for private firms to invest in new technologies.[20]

3. THE EFFECTIVENESS OF POLICY APPROACHES

The policy tools for reducing carbon dioxide emissions include pricing such emissions to discourage the use of fossil fuels and subsidizing the research and development of technologies that would reduce emissions. Those policy tools could be used individually or together (either simultaneously or sequentially). The empirical evidence available to evaluate policy options is limited, but that which is available can be used to inform simulation models. Those models can be used to compare the potential for R&D subsidies and carbon-pricing policies to increase net benefits from reducing carbon emissions or to evaluate the costs of achieving a given target for emissions. Those simulations suggest that the most efficient means of reducing carbon emissions would include applying both policy tools simultaneously; the gains from R&D alone would be unlikely to eliminate the value of near-term reductions in emissions and, thus, to obviate the benefits of near-term pricing.

Policy Simulations

The basic approach of models that simulate climate change policies is to measure the trade-off between consumption today and consumption in the future—where consumption is broadly defined to include the value of goods sold in markets, such as agricultural products, and the value of nonmarketed goods, such as ecosystems or longer life spans. Efforts to limit green-house gases in the near term would reduce the amount of resources that could be devoted to current consumption or to other investments. However, because those efforts are undertaken to reduce potential climate change damages in subsequent time periods, they lead to greater consumption in the future.

Policies are "efficient" if they maximize the present value of consumption over time. In other words, efficient policies would achieve the best balance between the reduction in current consumption that would result from today's carbon-reducing policies (policy costs) and the increase in consumption—

measured in present-value terms—that future generations might enjoy as a result of those policies (policy benefits). Alternatively, policies may efficiently meet a given emissions-reduction target if they do so with the lowest possible reduction in consumption over time.

Although simulation models are the best method currently available to evaluate policy options that address climate change, they are beset by uncertainties. Key uncertainties associated with measuring the costs of policy options include:

- The magnitude of emissions in the absence of policies—the lower those emissions are expected to be, the fewer reductions will be needed to meet a given target and the lower the costs will be;
- The costs of achieving a given reduction in emissions; and
- The effectiveness of policy tools in bringing about the lowest-cost reductions at a given point in time or in stimulating the development of new technologies.

Key uncertainties associated with measuring the benefits of policy options include:

- The effects of reductions in carbon emissions on the average global temperature;
- The distribution of reductions in global temperatures across seasons and regions, and the effects of those changes on other characteristics of climate, such as rainfall, severity of storms, and sea level;
- The effects of changes in regional climates on natural and human systems, such as agricultural crops, property, species, and human health; and
- The valuation of policy-induced reductions in damages to natural and human systems.[21]

Given the tremendous uncertainties involved in measuring policy costs and benefits, quantitative recommendations about the most efficient levels of prices or R&D subsidies are liable to be highly inaccurate. Qualitative policy recommendations about the appropriate timing, or relative importance, of policies are most robust if different modeling efforts reach similar conclusions or if a given model comes to the same qualitative conclusion when key assumptions are altered over the relevant range of potential values.

Results

The Congressional Budget Office identified three published analyses that simulate the effects of both emissions pricing and R&D.[22] While CBO recognizes the limitations of those modeling efforts and does not endorse any specific quantitative results, those models suggest that a combination of the two approaches—pricing emissions in the near term and funding R&D—would be necessary to reduce carbon emissions at the lowest possible cost. Further, they suggest that the largest gains in efficiency are likely to come from pricing emissions rather than from funding R&D. The simulation results described below reflect policies that are implemented at the global level.

One modeling effort concluded that the net benefits resulting from an R&D policy that balanced the costs and benefits of R&D efforts would be less than half of the net benefits that would result from an efficient carbon tax policy with no R&D subsidies.[23] Further, that effort found that the R&D policy would not reduce the benefits of near-term pricing of carbon because it would not reduce the value of near-term reductions in emissions. Two limitations of that modeling effort are that it did not examine the combined effects of the two policies, and it did not include an alternative energy technology (such as solar or wind generation); rather, R&D subsidies led to improvements in energy efficiency.[24]

A second researcher, using a model that included the possibility that R&D could improve both energy efficiency and an alternative energy technology, found that subsidizing R&D would provide only a small increase in net benefits relative to the maximum gains available if policy-makers enacted only an efficient carbon-pricing policy.[25] Further, the results suggested that subsidizing R&D would cause little change in the efficient price of emissions, or the amount of reductions, in the near term. To test the sensitivity of the findings to changes in assumptions, the researcher assumed that the additional R&D induced by subsidies would not crowd out any other research. That assumption did not change the researcher's qualitative results.[26]

A third modeling effort examined the ability of carbon pricing and R&D subsidies to minimize the cost of achieving a specific reduction in carbon emissions—in this case, 15 percent—rather than to maximize net benefits. That effort concluded that carbon pricing would achieve the target at a significantly lower cost than would an R&D subsidy for alternative energy sources. The cost of achieving the reductions with a carbon tax was roughly 11 percent of the cost of achieving the reductions with an R&D subsidy. Further, combining both policies would reduce the cost of achieving the emissions target by only 9 percent when compared with the cost of reaching it with only a carbon tax.[27]

Those models suggest that pricing emissions would contribute more to minimizing the cost of reducing emissions than would subsidizing investments in R&D. Given the uncertainties described above and the global nature of the simulation models, the quantitative results of such models should be viewed as very imprecise and corresponding conclusions about domestic policies should be made with caution. However, the qualitative finding that a cost-effective approach to reducing emissions would entail both funding R&D and pricing carbon in the near term is likely to be robust for several reasons:

- At any point in time, there is a cost continuum for emissions reductions, ranging from low-cost to high-cost opportunities. Unless R&D efforts were to virtually eliminate the value of near-term reductions in emissions, waiting to begin initial pricing (to encourage low-cost reductions) would increase the overall long-run cost of reducing emissions. Given reasonable assumptions about the costs of and gains from near-term R&D, the possibility that R&D would eliminate the value of near-term emissions reductions appears unlikely.
- Both pricing emissions and funding R&D would impose costs on the economy. Consequently, reducing emissions in the most cost-effective way would entail balancing the costs—and the expected payoffs—of both policies.
- Analyses that consider the costs and benefits of both carbon pricing and R&D all come to the same qualitative conclusion: near-term pricing of carbon emissions is an element of a cost-effective policy approach. That result holds even though studies make different assumptions about the availability of alternative energy technologies, the amount of crowding out caused by federal subsidies, and the form of the policy target (maximizing net benefits versus minimizing the cost of reaching a target).

The models described above are not well suited to account for the possibility that greenhouse gases could build up to a critical level, or threshold, in the atmosphere and thus could trigger a rapid increase in damages. Some analysts suggest the potential existence of such a threshold could call into question the value of making near-term reductions using existing technologies. In order to avoid passing such a threshold, it may be necessary to develop fundamentally new technologies that could provide a large share of the world's energy needs without releasing carbon or that could sequester similarly large shares of carbon emissions. Near-term pricing of emissions would impose costs on the economy but would not ultimately prevent any such threshold from being crossed—and large-scale damages from occurring. However, near-term pricing could have substantial value even under such circumstances. That value could stem from three sources:

- Near-term reductions in emissions could delay the crossing of a critical threshold and thus the point at which severe damages might occur. That delay has intrinsic value because the farther in the future that damages occur, the less value they have in the present.[28]
- Near-term reductions in emissions could allow time for fundamentally new technologies to be developed and put in place. That additional time could be essential to avoid crossing a critical threshold, given that the timing, and eventual outcomes, of R&D efforts are highly uncertain and that the process of putting new energy systems in place could be slow and costly.
- Near-term pricing could motivate current reductions in emissions whose cost is less than the present value of the expected (but uncertain) cost of reducing emissions using future technologies.

Limitations

Determining the appropriate mix of policies to address climate change is complicated by the fact that future policies would be layered on a complex mix of current and past policies, all of which affect the use of fossil fuels and their alternatives. Current policies subsidize the production and use of both fossil fuels and their alternatives. In addition, other existing federal, state, and local policies influence decisions about fossil fuels in less direct ways. While those existing policies serve other objectives, they also influence current and future carbon emissions.

The models that estimate the costs and benefits of pricing carbon emissions do not account for existing policies, and it is unclear how the quantitative conclusions of those models would be altered by their inclusion. The qualitative conclusion that a gradually increasing price on carbon emissions is likely to help minimize the costs of reducing carbon emissions is likely to be robust. No current policy attaches a cost to all fuels on the basis of their contribution to climate change; as a result, firms, households, governments, and other organizations do not face an incentive to take those costs into account in their production and consumption activities.

The picture for federal subsidies for R&D is also complex. New technologies are eligible for patents and are currently subsidized through general tax credits for R&D expenses. In addition, the federal government provides direct funding for some low-carbon energy sources, such as solar, nuclear, and wind power. Determining whether the magnitude of those subsidies, or the design and direction of the federal subsidy program, is appropriate is beyond the scope of this paper. To the extent that there are additional spillover benefits from R&D on carbon-reducing technologies, providing additional subsidies for R&D could be efficient.[29] Further, in the absence of carbon pricing, an R&D subsidy for carbon-reducing technologies could be justified by the external costs of carbon emissions; however, an R&D subsidy would offer a less direct method of reducing those external costs.

Administering a carbon-pricing program or initiating a larger R&D program for carbon-reducing technologies would entail costs. A full accounting of policy costs would entail weighing those administrative costs against the increase in net benefits that the policy would create. Available research indicates that the costs of implementing a carbon cap-and-trade program, or a tax on carbon emissions, are likely to vary significantly depending on where the cap, or tax, is placed. Placing the cap, or tax, upstream—on fossil fuel producers and importers— could minimize program costs. An upstream design could also provide incentives for firms to sequester carbon, as well as reduce their emissions, but the costs of implementation would probably be higher.[30]

Further, the models described above assume that policies are implemented in a perfect fashion. For example, they assume that a carbon tax would equalize the costs of reductions in emissions across all sources at a given point and would, in the absence of spillover benefits, lead to efficient investments in new technologies. In the case of R&D subsidies, they assume that federal subsidies would augment private R&D efforts up to the efficient point, with no distortions in the type of research that would be conducted, and that once the new technologies were developed, their cost savings would be fully captured. In reality, both policies would fall short of that theoretical ideal.

Finally, this study discusses policies designed to reduce carbon emissions in the United States. However, the causes and consequences of climate change are global, and the most cost-effective mitigation policies would require coordinated international efforts. Policies designed to promote low-cost reductions in emissions in rapidly developing countries, such as India and China, will be an important part of the policy mix for addressing climate change, though they are not addressed in this paper. If a domestic carbon-pricing program led to significant increases in the prices of U.S.-produced goods that were not matched by other countries, then carbon-intensive industries might choose to relocate to countries that do not have similar restrictions, diminishing the effectiveness of a U.S. carbon-pricing program. Adjusting the cap or tax to offset those effects could increase the cost of administering the program. Conversely, the establishment of a U.S. carbon-pricing program could affect the incentives of other countries to adopt similar restrictions. Moreover, successful domestic R&D efforts would lower the costs of reducing carbon emissions in other countries as well as within the United States. The extent to which those new technologies would be deployed at home and abroad would depend on whether incentives to reduce carbon emissions were put in place, or whether they could be cost-effective for other reasons, such as by providing energy at a lower cost than existing fossil fuel sources.

End Notes

[1] For a discussion of the uncertainty underlying estimates of the costs and benefits of reducing emissions and the implications for policy formation, see Congressional Budget Office, *Uncertainty in Analyzing Climate Change: Policy Implications* (January 2005).

[2] By discounting future damages to the present, the costs and benefits of undertaking actions that reduce climate change are placed on a common temporal footing. The discount rate chosen to compare avoided future damages (benefits) and costs is controversial. For a discussion, see Congressional Budget Office, *Uncertainty in Analyzing Climate Change* and *The Economics of Climate Change: A Primer* (April 2003).

[3] An exception to this would be if the net benefits of reducing carbon emissions were too small to justify the costs of administering either or both policies.

[4] Under an upstream system, fossil fuel producers and importers would pay the tax or purchase the allowance, and fossil fuel prices would increase accordingly. Those price increases would filter through the economy, increasing the prices of goods and services in proportion to the carbon emissions that their production and consumption generated.

[5] Electricity producers would have incentives to sequester carbon only if they were required to pay the tax directly or to hold the allowance for each ton of carbon they emitted. If the tax or allowance requirement was imposed upstream—on producers and importers of fossil fuels—electricity producers would have an incentive to minimize their use of carbon-intensive fuels, but they would not have an incentive to sequester carbon emissions unless special provisions were made. See Congressional Budget Office, *Issues in the Design of a Cap-and-Trade Program for Carbon Emissions*, CBO Issue Brief (November 25, 2003).

[6] Note that higher prices for fossil fuels resulting from an increasing scarcity of those fuels would also promote decreased use. However, fossil fuel use would still exceed the socially efficient amount because the prices would not reflect the external costs of carbon emissions.

[7] Studies have demonstrated the effectiveness of higher energy prices in stimulating R&D on renewable energy technologies. For example, see David Popp, "Induced Innovation and Energy Prices," *American Economic Review*, vol. 92 (2002), pp. 160-180.

[8] For a discussion of how near-term abatement can lower the cost of future carbon reductions by creating opportunities for learning by doing, see Michael Grubb, "Technologies, Energy Systems, and the Timing of CO_2 Abatement," *Energy Policy*, vol. 25, no. 2 (1997), pp. 159-172; Arnulf Grubler and Sabine Messner, "Technological Change and the Timing of Mitigation Measures," *Energy Economics*, vol. 20 (1998), pp. 495-512; and B.C.C. van der Zwann and others, "Endogenous Technological Change in Climate Change Modeling," *Energy Economics,* vol. 24 (2002), pp. 1-19.

[9] That measure of expected damages depends on many uncertain factors, including the time that carbon lingers in the atmosphere, the path of emissions in the future, the change in climate associated with higher atmospheric stocks of carbon, the physical damages associated with changes in climate, and the value of those damages. For a discussion of those factors, see Congressional Budget Office, *Uncertainties in Analyzing Climate Change: Policy Implications* (January 2005).

[10] As a first approximation, intertemporal efficiency would require that carbon prices rise at the real (inflation-adjusted) rate of interest minus the rate at which carbon was absorbed into the ocean and thus disappeared from the atmosphere. See William D. Nordhaus, *Life After Kyoto: Alternative Approaches to Global Warming Policies*, Working Paper No. 11889 (Cambridge, Mass.: National Bureau of Economic Research, December 2005), p. 9. Because damages stem from the accumulation of carbon in the atmosphere, they are expected to be greater in the future than in the present. The efficient price of carbon increases over time to represent the growing weight placed on future damages as they become closer in time.

[11] The baseline capital stock could become more or less carbon intensive as the relative prices of fossil fuels and alternative technologies changed over time. However, in the absence of carbon pricing—or in anticipation of it—there would be no incentive for firms or households to make choices on the basis of the carbon emissions that resulted from the different technologies. As a result, even models that suggest that a stated target for the atmospheric concentration of carbon could be met with little divergence from baseline emissions in the near term find that some near-term reductions would have value (this is reflected by a positive "shadow price" for such reductions). For an example, see Richard Richels and Jae Edmonds, "The Economics of Stabilizing Atmospheric CO_2 Concentrations," *Energy Policy*, vol. 23, no. 4/5 (1995), pp. 373-378. Information about the shadow price of carbon was based on personal communication with the lead author.

[12] Some models incorporate information about specific alternative energy technologies, including their initial price and the prospects that their costs would fall. Those models—called bottom-up models because of the detail that they contain on specific technologies—tend to be relatively optimistic about the prospects for technological advancement but typically fail to represent the costs that might be associated with research efforts. Top-down models, in contrast, tend to have better information on the links between environmental policy and macroeconomic performance (better capturing the costs of R&D efforts), but they have considerably less detail about specific alternative energy technologies and tend to be less optimistic about the prospects of cost reductions. For a more complete discussion of the strengths and limitations of top-down and bottom-up models, see Leon E. Clarke and John P. Weyant, "Modeling Induced Technological Change: An Overview," in Arnulf Grubler, Nebojsa Nakiconovic, and William D. Nordhaus, eds., *Technological Change and the Environment* (Washington, D.C.: Resources for the Future, 2002), pp. 343-349; and David Popp, "Entice-BR: The Effects of Backstop Technology and R&D on Climate Policy Models," *Energy Economics,* vol. 28 (2006), pp. 189-191.

[13] Carbon emissions would be roughly 80 percent higher if a given amount of heat was generated by coal (or 50 percent higher if generated by fuel oil) rather than by natural gas. See www.epa.gov/air market/epa-ipm/chapter8.pdf.

[14] For a more complete description of these and other ideas, see Kenneth Caldeira and others, *Climate Change Technology Exploratory Research* (Washington, D.C.: Climate Policy Center, December 2005).

[15] Estimates of social rates of return (society's benefits) resulting from private investments in research and development are highly uncertain, but some researchers suggest that they average as much as 30 percent to 50 percent. In contrast, private marginal rates of return are estimated to be between 7 percent and 15 percent. See research cited in David Popp, *R&D Subsidies and Climate Policy: Is There a 'Free Lunch'?* Working Paper No. 10880 (Cambridge, Mass.: National Bureau of Economic Research, October 2004), p. 4; and Edward Mansfield, "Macroeconomic Policy and Technological Change," in J.C. Fuhrere and J. Sneddon Little, eds., *Technology and Growth, Conference Proceedings* (Boston, Mass.: Federal Reserve Bank of Boston, 1996), p. 191.

[16] See Austin Goolsbee, "Does Government R&D Policy Mainly Benefit Scientists and Engineers?" *American Economic Review,* vol. 88 (1988), pp. 298-302.

[17] For example, one study concluded that incremental improvements in existing technologies could limit carbon emissions over the next 50 years to a trajectory that would avoid a doubling of the pre-industrial concentration. See S. Pacala and R. Scolow, "Stabilization Wedges: Solving the Climate Problem for the Next 50 Years with Current Technologies," *Science*, vol. 305 (August 13, 2004). In contrast, another study concluded that currently known technologies have severe deficiencies that limit their ability to stabilize the global climate. See Martin I. Hoffert and others, "Advanced Technology Paths to Global Climate Stability: Energy for a Green-house Planet," *Science*, vol. 298 (November 1, 2002).

[18] See Congressional Research Service, *Tax Expenditures: Compendium of Background Material on Individual Provisions* (S.PRT. 108-54), pp. 59-70.

[19] Since 1981, when this tax credit was first enacted, it has been extended 10 times, most recently through December 31, 2005. Although the statutory rate of the credit is 20 percent, the actual cost reduction is less than 20 percent (13 percent for some expenditures and 6.5 percent for others) because of certain rules governing how the credit is computed.

[20] In the absence of carbon pricing, the research subsidy would simultaneously reduce the distortions in research incentives caused by two market failures. See Lawrence H. Goulder and Stephen H. Schneider, "Induced Technological Change and the Attractiveness of CO_2 Abatement Policies," *Resource and Energy Economics*, vol. 21 (1999), pp. 211-253.

[21] See Congressional Budget Office, *Uncertainty in Analyzing Climate Change: Policy Implications* (January 2005), for a more detailed discussion of the uncertainties associated with the costs and benefits of climate policies.

[22] Those models assume that carbon pricing would stimulate less private investment in R&D than would be best for society given the existence of spillover benefits. Further, they assume that public funds would subsidize private R&D efforts. Their conclusions, therefore, do not stem from assumptions about the success or failure of federal R&D programs.

[23] R&D benefits were assumed to include both spillover benefits and the reduction in external costs that would result from the new technologies.

[24] That modeling effort also estimated that an efficient tax on carbon emissions would lead to about twice the level of reductions in emissions that would result from an efficient R&D subsidy. See William D. Nordhaus, "*Modeling Induced Innovation in Climate Change Policy*," in A. Grubler, N. Nakicenovic, and W.D. Nordhaus, eds., Modeling Induced Innovation in Climate Change Policy (Washington, D.C.: Resources for the Future Press, 2002). Nordhaus assumes that technology improvements are exogenous in the carbon-tax-only case.

[25] Net benefits increased by 7 percent when R&D subsidies were added to a carbon-pricing policy. See David Popp, "Entice-BR: The Effects of Backstop Technology and R&D on

Climate Policy Models," *Energy Economics*, vol. 28 (2006), pp. 188-222. Popp does not report the change in efficient prices and emissions reductions under the sensitivity analysis. Another analysis (which did not include a backstop technology) by Popp found that the net benefits associated with an efficient R&D subsidy were only 11 percent of the maximum gains available with an efficient carbon-pricing policy and an efficient R&D subsidy. In contrast, the net benefits associated with the carbon-tax-only case were 95 percent of the potential maximum gains. To test the robustness of his results, Popp assumed that the returns to energy R&D were twice as high as the returns to other forms of R&D, and under that assumption he still found that the net benefits of an efficient R&D strategy would amount to less than 55 percent of the potential gains from the combined pricing/R&D subsidy approach. See David Popp, *R&D Subsidies and Climate Policy: Is There a 'Free Lunch'?* Working Paper No. 10880 (Cambridge, Mass.: National Bureau of Economic Research, October 2004).

[26] Under that assumption, Popp found that an R&D subsidy would increase potential net benefits by an additional 30 percent.

[27] See Stephen H. Schneider and Lawrence H. Goulder, "Achieving Low-Cost Emissions Targets," *Nature*, vol. 489 (September 1997), pp. 13-14. In addition, those researchers demonstrated that the merits of a tax and of R&D subsidies are sensitive to the magnitude of prior distortions created by existing subsidies to R&D for alternative and fossil-fuel-based energy. A carbon tax can help undo, or exacerbate, an existing imbalance between incentives for R&D on alternative and fossil fuels. Further, the desirability of subsidizing R&D on alternative energy technologies depends not only on the existence of knowledge spillovers from such research, but on the magnitude of those spillover benefits relative to spill-over benefits from R&D in other sectors and on whether a carbon tax is in place. See Lawrence H. Goulder and Stephen H. Schneider, "Induced Technological Change and the Attractiveness of CO_2 Abatement Policies," *Resource and Energy Economics*, vol. 21 (1999), pp. 211-253.

[28] Future damages are discounted to reflect the fact that, in a growing economy, future generations would have more resources with which to address such damages.

[29] In the absence of additional R&D subsidies, relatively large spill-over benefits from R&D on carbon-reducing technologies could also reduce the economic costs of a carbon tax. By redirecting research toward those technologies, the carbon tax would both internalize the costs of carbon emissions and help reduce the under-funding of private research efforts (due to unaddressed spillover benefits) in that area. For a discussion of that point, see Goulder and Schneider, "Induced Technological Change and the Attractiveness of CO_2 Abatement Policies." Further, if spillover benefits (what remained after accounting for existing subsidies) are small for R&D on carbon-reducing technologies (relative to other types of R&D), then providing additional subsidies for R&D on carbon-reducing technologies could be inefficient in that it could crowd out higher-valued research in other areas.

[30] For a discussion of implementation issues, see Congressional Budget Office, *An Evaluation of Cap-and-Trade Programs for Reducing U.S. Carbon Emissions* (June 2001) and CBO's Comments on the White Paper "Design Elements of a Mandatory Market-Based Greenhouse Gas Regulatory System" (March 12, 2006).

CHAPTER SOURCES

The following chapters have been previously published:

Chapter 1 – This is an edited, excerpted and augmented edition of a United States Congressional Budget Office, Economic and Budget Issue Brief, dated October 6, 2008.

Chapter 2 – These remarks were delivered as Statement of Peter R. Orszag, Director, United States Congressional Budget Office, before the Committee on Ways and Means, U.S. House of Representatives, dated September 18, 2008.

Chapter 3 – This is an edited, excerpted and augmented edition of a United States Congressional Budget Office Study, dated February 2008.

Chapter 4 - This is an edited, excerpted and augmented edition of a United States Congressional Budget Office paper, dated September 2006.

INDEX

D

E

F

G

S

T

U

V

W

Y